AF863001

Springer Proceedings in Mathematics & Statistics

Volume 356

This book series features volumes composed of selected contributions from workshops and conferences in all areas of current research in mathematics and statistics, including operation research and optimization. In addition to an overall evaluation of the interest, scientific quality, and timeliness of each proposal at the hands of the publisher, individual contributions are all refereed to the high quality standards of leading journals in the field. Thus, this series provides the research community with well-edited, authoritative reports on developments in the most exciting areas of mathematical and statistical research today.

More information about this series at http://www.springer.com/series/10533

Benjamin Ong · Jacob Schroder · Jemma Shipton ·
Stephanie Friedhoff
Editors

Parallel-in-Time Integration Methods

9th Parallel-in-Time Workshop, June 8–12, 2020

Springer

Editors
Benjamin Ong
Department of Mathematical Sciences
Michigan Technological University
Houghton, MI, USA

Jacob Schroder
Department of Mathematics and Statistics
University of New Mexico
Albuquerque, NM, USA

Jemma Shipton
Department of Mathematics
University of Exeter
Exeter, UK

Stephanie Friedhoff
Department of Mathematics
University of Wuppertal
Wuppertal, Nordrhein-Westfalen, Germany

ISSN 2194-1009 ISSN 2194-1017 (electronic)
Springer Proceedings in Mathematics & Statistics
ISBN 978-3-030-75932-2 ISBN 978-3-030-75933-9 (eBook)
https://doi.org/10.1007/978-3-030-75933-9

Mathematics Subject Classification: 65Y05, 65Y20, 65L06, 65M12, 65M55

© The Editor(s) (if applicable) and The Author(s), under exclusive license to Springer Nature Switzerland AG 2021
This work is subject to copyright. All rights are solely and exclusively licensed by the Publisher, whether the whole or part of the material is concerned, specifically the rights of translation, reprinting, reuse of illustrations, recitation, broadcasting, reproduction on microfilms or in any other physical way, and transmission or information storage and retrieval, electronic adaptation, computer software, or by similar or dissimilar methodology now known or hereafter developed.
The use of general descriptive names, registered names, trademarks, service marks, etc. in this publication does not imply, even in the absence of a specific statement, that such names are exempt from the relevant protective laws and regulations and therefore free for general use.
The publisher, the authors and the editors are safe to assume that the advice and information in this book are believed to be true and accurate at the date of publication. Neither the publisher nor the authors or the editors give a warranty, expressed or implied, with respect to the material contained herein or for any errors or omissions that may have been made. The publisher remains neutral with regard to jurisdictional claims in published maps and institutional affiliations.

This Springer imprint is published by the registered company Springer Nature Switzerland AG
The registered company address is: Gewerbestrasse 11, 6330 Cham, Switzerland

Preface

This volume includes contributions from the 9th Parallel-in-Time (PinT) workshop, held virtually during June 8–12, 2020, due to the COVID-19 pandemic. Over 100 researchers participated in the 9th PinT Workshop, organized by the editors of this volume. The PinT workshop series is an annual gathering devoted to the field of time-parallel methods, aiming to adapt existing computer models to next-generation machines by adding a new dimension of scalability. As the latest supercomputers advance in microprocessing ability, they require new mathematical algorithms in order to fully realize their potential for complex systems. The use of parallel-in-time methods will provide dramatically faster simulations in many important areas, including biomedical (e.g., heart modeling), computational fluid dynamics (e.g., aerodynamics and weather prediction), and machine learning applications. Computational and applied mathematics is crucial to this progress, as it requires advanced methodologies from the theory of partial differential equations in a functional analytic setting, numerical discretization and integration, convergence analyses of iterative methods, and the development and implementation of new parallel algorithms. The workshop brings together an interdisciplinary group of experts across these fields to disseminate cutting-edge research and facilitate discussions on parallel time integration methods.

Houghton, MI, USA — Benjamin Ong
Albuquerque, NM, USA — Jacob Schroder
Exeter, UK — Jemma Shipton
Wuppertal, Germany — Stephanie Friedhoff

Contents

Contributors

Tommaso Buvoli University of California, Merced, Merced, CA, USA

Federico Danieli University of Oxford, Oxford, England

Andreas Gocht Center of Information Services and High Performance Computing, Dresden, Germany

Sebastian Götschel Chair Computational Mathematics, Institute of Mathematics, Hamburg University of Technology, Hamburg, Germany

Andreas Hessenthaler University of Oxford, Oxford, England

Michael Knobloch Forschungszentrum Jülich GmbH, Jülich Supercomputing Centre, Jülich, Germany

Sebastian Lührs Forschungszentrum Jülich GmbH, Jülich Supercomputing Centre, Jülich, Germany

Michael Minion Lawrence Berkeley National Laboratory, Berkeley, CA, USA

Wayne Mitchell Heidelberg University, Heidelberg, Germany

Daniel Ruprecht Chair Computational Mathematics, Institute of Mathematics, Hamburg University of Technology, Hamburg, Germany

Ben S. Southworth Los Alamos National Laboratory, Los Alamos, NM, USA

Robert Speck Jülich Supercomputing Centre, Forschungszentrum Jülich GmbH, Jülich, Germany

Raymond van Venetië Korteweg–de Vries (KdV) Institute for Mathematics, University of Amsterdam, Amsterdam, The Netherlands

Jan Westerdiep Korteweg–de Vries (KdV) Institute for Mathematics, University of Amsterdam, Amsterdam, The Netherlands

Tight Two-Level Convergence of Linear Parareal and MGRIT: Extensions and Implications in Practice

Ben S. Southworth, Wayne Mitchell, Andreas Hessenthaler, and Federico Danieli

Abstract Two of the most popular parallel-in-time methods are Parareal and multigrid-reduction-in-time (MGRIT). Recently, a general convergence theory was developed in Southworth [17] for linear two-level MGRIT/Parareal that provides necessary and sufficient conditions for convergence, with tight bounds on worst-case convergence factors. This paper starts by providing a new and simplified analysis of linear error and residual propagation of Parareal, wherein the norm of error or residual propagation is given by one over the minimum singular value of a certain block bidiagonal operator. New discussion is then provided on the resulting necessary and sufficient conditions for convergence that arise by appealing to block Toeplitz theory as in Southworth [17]. Practical applications of the theory are discussed, and the convergence bounds demonstrated to predict convergence in practice to high accuracy on two standard linear hyperbolic PDEs: the advection(-diffusion) equation and the wave equation in first-order form.

1 Background

Two of the most popular parallel-in-time methods are Parareal [10] and multigrid-reduction-in-time (MGRIT) [5]. Convergence of Parareal/two-level MGRIT has been considered in a number of papers [1, 4, 6–9, 14, 18, 19]. Recently, a general convergence theory was developed for linear two-level MGRIT/Parareal that provides necessary and sufficient conditions for convergence, with tight bounds on worst-case convergence factors [17], and does not rely on assumptions of diagonalizability of the underlying operators. Section 2 provides a simplified analysis of linear Parareal and MGRIT that expresses the norm of error or residual propagation of two-level

B. S. Southworth (✉)
Los Alamos National Laboratory, Los Alamos, NM 87544, USA
e-mail: southworth@lanl.gov

W. Mitchell
Heidelberg University, Heidelberg, Germany

A. Hessenthaler · F. Danieli
University of Oxford, Oxford, England

© The Author(s), under exclusive license to Springer Nature Switzerland AG 2021
B. Ong et al. (eds.), *Parallel-in-Time Integration Methods*, Springer Proceedings in Mathematics & Statistics 356, https://doi.org/10.1007/978-3-030-75933-9_1

linear Parareal and MGRIT precisely as one over the minimum singular value of a certain block bidiagonal operator (rather than the more involved pseudoinverse approach used in [17]). We then provide a new discussion on the resulting necessary and sufficient conditions for convergence that arise by appealing to block Toeplitz theory [17]. In this paper, we define *convergence* after a given number of iterations, as a guarantee that the ℓ^2-norm of the error or residual will be reduced, regardless of right-hand side or initial guess. In addition, the framework developed in [17] (which focuses on, equivalently, convergence of error/residual on C-points or convergence of error/residual on *all* points for two or more iterations) is extended to provide necessary conditions for the convergence of a single iteration on all points, followed by a discussion putting this in the context of multilevel convergence in Sect. 2.4. Practical applications of the theory are discussed in Sect. 3, and the convergence bounds demonstrated to predict convergence in practice to high accuracy on two standard linear hyperbolic PDEs: the advection(-diffusion) equation and the wave equation in first-order form.

2 Two-Level Convergence

2.1 A Linear Algebra Framework

Consider time integration with N discrete time points. Let $\Phi(t)$ be a time-dependent, linear, and stable time propagation operator, with subscript ℓ denoting $\Phi_\ell := \Phi(t_\ell)$, and let $\mathbf{u}_\ell$ denote the (spatial) solution at time point t_ℓ. Then, consider the resulting space-time linear system,

$$A\mathbf{u} := \begin{bmatrix} I & & & & \\ -\Phi_1 & I & & & \\ & -\Phi_2 & I & & \\ & & \ddots & \ddots & \\ & & & -\Phi_{N-1} & I \end{bmatrix} \begin{bmatrix} \mathbf{u}_0 \\ \mathbf{u}_1 \\ \mathbf{u}_2 \\ \vdots \\ \mathbf{u}_{N-1} \end{bmatrix} = \mathbf{f}. \tag{1}$$

Clearly, (1) can be solved directly using block forward substitution, which corresponds to standard sequential time stepping. Linear Parareal and MGRIT are reduction-based multigrid methods, which solve (1) in a parallel iterative manner.

First, note there is a closed form inverse for matrices with the form in (1), which will prove useful for further derivations. Excusing the slight abuse of notation, define $\Phi_i^j := \Phi_i \Phi_{i-1} ... \Phi_j$. Then,

$$\begin{bmatrix} I & & & & \\ -\Phi_1 & I & & & \\ & -\Phi_2 & I & & \\ & & \ddots & \ddots & \\ & & & -\Phi_{N-1} & I \end{bmatrix}^{-1} = \begin{bmatrix} I & & & & & \\ \Phi_1 & I & & & & \\ \Phi_2^1 & \Phi_2 & I & & & \\ \Phi_3^1 & \Phi_3^2 & \Phi_3 & I & & \\ \vdots & \vdots & & \ddots & \ddots & \\ \Phi_{N-1}^1 & \Phi_{N-1}^2 & \cdots & \cdots & \Phi_{N-1} & I \end{bmatrix}. \tag{2}$$

Now suppose we coarsen in time by a factor of k, that is, for every kth time point, we denote a C-point, and the $k-1$ points between each C-point are considered F-points (it is not necessary that k be fixed across the domain, rather this is a simplifying assumption for derivations and notation; for additional background on C-points, F-points, and the multigrid-in-time framework, see [5]). Then, using the inverse in (2) and analogous matrix derivations as in [17], we can eliminate F-points from (1) and arrive at a Schur complement of A over C-points, given by

$$A_\Delta := \begin{bmatrix} I & & & & \\ -\Phi_k^1 & I & & & \\ & -\Phi_{2k}^{k+1} & I & & \\ & & \ddots & \ddots & \\ & & & -\Phi_{(N_c-1)k}^{(N_c-2)k+1} & I \end{bmatrix}. \tag{3}$$

Notice that the Schur complement coarse-grid operator in the time-dependent case does exactly what it does in the time-independent case: it takes k steps on the fine grid, in this case using the appropriate sequence of time-dependent operators.

A Schur complement arises naturally in reduction-based methods when we eliminate certain degrees-of-freedom (DOFs). In this case, even computing the action of the Schur complement (3) is too expensive to be considered tractable. Thus, parallel-in-time integration is based on a Schur complement approximation, where we let Ψ_i denote a non-Galerkin approximation to $\Phi_{ik}^{(i-1)k+1}$ and define the "coarse-grid" time integration operator $B_\Delta \approx A_\Delta$ as

$$B_\Delta := \begin{bmatrix} I & & & & \\ -\Psi_1 & I & & & \\ & -\Psi_2 & I & & \\ & & \ddots & \ddots & \\ & & & -\Psi_{N_c-1} & I \end{bmatrix}.$$

Convergence of iterative methods is typically considered by analyzing the error and residual propagation operators, say $\mathcal{E}$ and $\mathcal{R}$. Necessary and sufficient conditions for convergence of an iterative method are that $\mathcal{E}^p, \mathcal{R}^p \to 0$ with p. In this case, eigenvalues of $\mathcal{E}$ and $\mathcal{R}$ are a poor measure of convergence, so we consider the ℓ^2-norm. For notation, assume we block partition A (1) into C-points and F-points and reorder $A \mapsto \begin{bmatrix} A_{ff} & A_{fc} \\ A_{cf} & A_{cc} \end{bmatrix}$. MGRIT is typically based on F- and FCF-relaxation; F-relaxation corresponds to sequential time stepping along contiguous F-points, while FCF-relaxation then takes a time step from the final F-point in each group to its adjacent C-point, and then repeats F-relaxation with the updated initial C-point value (see [5]). Letting subscript F denote F-relaxation and subscript FCF denote FCF-relaxation, error and residual propagation operators for two-level Parareal/MGRIT are derived in [17] to be

$$\begin{aligned}
\mathcal{E}_F^p &:= \begin{bmatrix} -A_{ff}^{-1}A_{fc} \\ I \end{bmatrix} \begin{bmatrix} \mathbf{0} & (I - B_\Delta^{-1}A_\Delta)^p \end{bmatrix}, \\
\mathcal{E}_{FCF}^p &:= \begin{bmatrix} -A_{ff}^{-1}A_{fc} \\ I \end{bmatrix} \begin{bmatrix} \mathbf{0} & \left((I - B_\Delta^{-1}A_\Delta)(I - A_\Delta)\right)^p \end{bmatrix}, \\
\mathcal{R}_F^p &:= \begin{bmatrix} \mathbf{0} \\ (I - A_\Delta B_\Delta^{-1})^p \end{bmatrix} \begin{bmatrix} -A_{cf}A_{ff}^{-1} & I \end{bmatrix}, \\
\mathcal{R}_{FCF}^p &:= \begin{bmatrix} \mathbf{0} \\ \left((I - A_\Delta B_\Delta^{-1})(I - A_\Delta)\right)^p \end{bmatrix} \begin{bmatrix} -A_{cf}A_{ff}^{-1} & I \end{bmatrix}.
\end{aligned} \tag{4}$$

In [17], the leading terms involving $A_{cf}A_{ff}^{-1}$ and $A_{ff}^{-1}A_{fc}$ (see [17] for representation in Φ) are shown to be bounded in norm $\leq k$. Thus, as we iterate $p > 1$, convergence of error and residual propagation operators is defined by iterations on the coarse space, e.g. $(I - B_\Delta^{-1}A_\Delta)$ for $\mathcal{E}_F$. To that end, define

$$\begin{aligned}
\widetilde{\mathcal{E}}_F &:= I - B_\Delta^{-1}A_\Delta, & \widetilde{\mathcal{E}}_{FCF} &:= (I - B_\Delta^{-1}A_\Delta)(I - A_\Delta), \\
\widetilde{\mathcal{R}}_F &:= I - A_\Delta B_\Delta^{-1}, & \widetilde{\mathcal{R}}_{FCF} &:= (I - A_\Delta B_\Delta^{-1})(I - A_\Delta).
\end{aligned} \tag{5}$$

2.2 A Closed Form for $\|\widetilde{\mathcal{E}}\|$ and $\|\widetilde{\mathcal{R}}\|$

F-relaxation: Define the shift operators and block diagonal matrix

$$I_L := \begin{bmatrix} \mathbf{0} & & & \\ I & \mathbf{0} & & \\ & \ddots & \ddots & \\ & & I & \mathbf{0} \end{bmatrix}, \quad I_z := \begin{bmatrix} I & & & \\ & \ddots & & \\ & & I & \\ & & & \mathbf{0} \end{bmatrix},$$

$$D := \begin{bmatrix} \Phi_k^1 - \Psi_1 & & & \\ & \ddots & & \\ & & \Phi_{(N_c-1)k}^{(N_c-2)k+1} - \Psi_{N_c-1} & \\ & & & \mathbf{0} \end{bmatrix}$$

and notice that $I_L D = (B_\Delta - A_\Delta)$ and $I_L^T I_L = I_z$. Further define $\widetilde{D}$ and $\widetilde{B}_\Delta$ as the leading principle submatrices of D and B_Δ, respectively, obtained by eliminating the last (block) row and column, corresponding to the final coarse-grid time step, $N_c - 1$.

Now note that $\widetilde{\mathcal{R}}_F = I - A_\Delta B_\Delta^{-1} = (B_\Delta - A_\Delta)B_\Delta^{-1} = I_L D B_\Delta^{-1}$ and $I_L^T I_L = I_z$. Then,

$$\|\widetilde{\mathcal{R}}_F\|^2 = \sup_{\mathbf{x}\neq\mathbf{0}} \frac{\langle I_L D B_\Delta^{-1}\mathbf{x}, I_L D B_\Delta^{-1}\mathbf{x}\rangle}{\langle \mathbf{x}, \mathbf{x}\rangle} = \sup_{\mathbf{x}\neq\mathbf{0}} \frac{\langle I_z D B_\Delta^{-1}\mathbf{x}, I_z D B_\Delta^{-1}\mathbf{x}\rangle}{\langle \mathbf{x}, \mathbf{x}\rangle}. \quad (6)$$

Because DB_Δ^{-1} is lower triangular, setting the last row to zero by multiplying by I_z also sets the last column to zero, in which case $I_z D B_\Delta^{-1} = I_z D B_\Delta^{-1} I_z$. Continuing from (6), we have

$$\|\widetilde{\mathcal{R}}_F\|^2 = \sup_{\mathbf{y}\neq\mathbf{0}} \frac{\langle \widetilde{D}\widetilde{B}_\Delta^{-1}\mathbf{y}, \widetilde{D}\widetilde{B}_\Delta^{-1}\mathbf{y}\rangle}{\langle \mathbf{y}, \mathbf{y}\rangle} = \|\widetilde{D}\widetilde{B}_\Delta^{-1}\|^2,$$

where $\mathbf{y}$ is defined on the lower dimensional space corresponding to the operators $\widetilde{D}$ and $\widetilde{B}_\Delta$. Recalling that the ℓ^2-norm is defined by $\|A\| = \sigma_{\max}(A) = 1/\sigma_{\min}(A^{-1})$, for maximum and minimum singular values, respectively,

$$\|\widetilde{\mathcal{R}}_F\| = \sigma_{\max}\left(\widetilde{D}\widetilde{B}_\Delta^{-1}\right) = \max_{\mathbf{x}\neq\mathbf{0}} \frac{\|\widetilde{D}\mathbf{x}\|}{\|\widetilde{B}_\Delta\mathbf{x}\|} = \frac{1}{\sigma_{\min}\left(\widetilde{B}_\Delta\widetilde{D}^{-1}\right)},$$

where

$$\widetilde{B}_\Delta\widetilde{D}^{-1} := \begin{bmatrix} I & & & \\ -\Psi_1 & I & & \\ & \ddots & \ddots & \\ & & -\Psi_{N_c-2} & I \end{bmatrix} \begin{bmatrix} \left(\Phi_k^1 - \Psi_1\right)^{-1} & & \\ & \ddots & \\ & & \left(\Phi_{(N_c-1)k}^{(N_c-2)k+1} - \Psi_{N_c-1}\right)^{-1} \end{bmatrix}.$$

Similarly, $\widetilde{\mathcal{E}}_F = I - B_\Delta^{-1}A_\Delta = B_\Delta^{-1} I_L D$. Define $\widehat{B}_\Delta$ as the principle submatrix of B_Δ obtained by eliminating the *first* row and column. Similar arguments as above then yield

$$\|\widetilde{\mathcal{E}}_F\| = \sigma_{\max}\left(\widehat{B}_\Delta^{-1}\widetilde{D}\right) = \max_{\mathbf{x}\neq\mathbf{0}} \frac{\|\widehat{B}_\Delta^{-1}\mathbf{x}\|}{\|\widetilde{D}^{-1}\mathbf{x}\|} = \frac{1}{\sigma_{\min}\left(\widetilde{D}^{-1}\widehat{B}_\Delta\right)},$$

where

$$\widetilde{D}^{-1}\widehat{B}_\Delta := \begin{bmatrix} \left(\Phi_k^1 - \Psi_1\right)^{-1} & & \\ & \ddots & \\ & & \left(\Phi_{(N_c-1)k}^{(N_c-2)k+1} - \Psi_{N_c-1}\right)^{-1} \end{bmatrix} \begin{bmatrix} I & & & \\ -\Psi_2 & I & & \\ & \ddots & \ddots & \\ & & -\Psi_{N_c-1} & I \end{bmatrix}.$$

FCF-relaxation: Adding the effects of (pre)FCF-relaxation, $\widetilde{\mathcal{E}}_{FCF} = B_\Delta^{-1}(B_\Delta - A_\Delta)(I - A_\Delta)$, where $(B_\Delta - A_\Delta)(I - A_\Delta)$ is given by the block diagonal matrix

$$= I_L^2 \begin{bmatrix} \left(\Phi_{2k}^{k+1} - \Psi_2\right)\Phi_k^1 & & & \\ & \ddots & & \\ & & \left(\Phi_{(N_c-1)k}^{(N_c-2)k+1} - \Psi_{N_c-1}\right)\Phi_{(N_c-2)k}^{(N_c-3)k+1} & & \\ & & & \mathbf{0} & \\ & & & & \mathbf{0} \end{bmatrix}.$$

Again analogous arguments as for F-relaxation can pose this as a problem on a nonsingular matrix of reduced dimensions. Let

$$\widetilde{D}_{fcf}^{-1}\overline{B}_\Delta := \begin{bmatrix} \left(\Phi_{2k}^{k+1} - \Psi_2\right)\Phi_k^1 & & \\ & \ddots & \\ & & \left(\Phi_{(N_c-1)k}^{(N_c-2)k+1} - \Psi_{N_c-1}\right)\Phi_{(N_c-2)k}^{(N_c-3)k+1} \end{bmatrix}^{-1} \begin{bmatrix} I & & & \\ -\Psi_3 & I & & \\ & \ddots & \ddots & \\ & & -\Psi_{N_c-1} & I \end{bmatrix}.$$

Then

$$\|\widetilde{\mathcal{E}}_{FCF}\| = \sigma_{\max}\left(\overline{B}_\Delta^{-1}\widetilde{D}_{fcf}\right) = \max_{\mathbf{x}\neq\mathbf{0}} \frac{\|\overline{B}_\Delta^{-1}\mathbf{x}\|}{\|\widetilde{D}_{fcf}^{-1}\mathbf{x}\|} = \frac{1}{\sigma_{\min}\left(\widetilde{D}_{fcf}^{-1}\overline{B}_\Delta\right)}. \tag{7}$$

The case of FCF-relaxation with residual propagation produces operators with a more complicated structure. If we simplify by assuming that Φ and Ψ commute, or consider pre-FCF-relaxation, rather than post-FCF-relaxation, analogous results to (7) follow a similar analysis, with the order of the B- and D-terms swapped, as was the case for F-relaxation.

2.3 *Convergence and the Temporal Approximation Property*

Now assume that $\Phi_i = \Phi_j$ and $\Psi_i = \Psi_j$ for all i, j, that is, Φ and Ψ are independent of time. Then, the block matrices derived in the previous section defining convergence of two-level MGRIT all fall under the category of block-Toeplitz matrices. By appealing to block-Toeplitz theory, in [17], tight bounds are derived on the appropriate minimum and maximum singular values appearing in the previous section, which are exact as the number of coarse-grid time steps $\to \infty$. The fundamental concept is the "temporal approximation property" (TAP), as introduced below, which provides necessary and sufficient conditions for two-level convergence of Parareal and MGRIT. Moreover, the constant with which the TAP is satisfied, φ_F or φ_{FCF}, provides a tight upper bound on convergence factors, that is asymptotically exact as $N_c \to \infty$. We present a simplified/condensed version of the theoretical results derived in [17] in Theorem 1.

Definition 1 (*Temporal approximation property*) Let Φ denote a fine-grid time-stepping operator and Ψ denote a coarse-grid time-stepping operator, for all time points, with coarsening factor k. Then, Φ satisfies an F-relaxation temporal approximation property (F-TAP), with respect to Ψ, with constant φ_F, if, for all vectors $\mathbf{v}$,

$$\|(\Psi - \Phi^k)\mathbf{v}\| \leq \varphi_F \left[\min_{x\in[0,2\pi]} \left\|(I - e^{\mathrm{i}x}\Psi)\mathbf{v}\right\| \right]. \tag{8}$$

Similarly, Φ satisfies an FCF-relaxation temporal approximation property (FCF-TAP), with respect to Ψ, with constant φ_{FCF}, if, for all vectors $\mathbf{v}$,

$$\|(\Psi - \Phi^k)\mathbf{v}\| \leq \varphi_{FCF} \left[\min_{x\in[0,2\pi]} \left\|(\Phi^{-k}(I - e^{\mathrm{i}x}\Psi))\mathbf{v}\right\| \right]. \tag{9}$$

Theorem 1 (Necessary and sufficient conditions) *Suppose Φ and Ψ are linear, stable ($\|\Phi^p\|, \|\Psi^p\| < 1$ for some p), and independent of time; and that $(\Psi - \Phi^k)$ is invertible. Further suppose that Φ satisfies an F-TAP with respect to Ψ, with constant φ_F, and Φ satisfies an FCF-TAP with respect to Ψ, with constant φ_{FCF}. Let $\mathbf{r}_i$ denote the residual after i iterations. Then, worst-case convergence of residual is bounded above and below by*

$$\frac{\varphi_F}{1 + O(1/\sqrt{N_c - i})} \leq \frac{\|\mathbf{r}_{i+1}^{(F)}\|}{\|\mathbf{r}_i^{(F)}\|} < \varphi_F,$$

$$\frac{\varphi_{FCF}}{1 + O(1/\sqrt{N_c - i})} \leq \frac{\|\mathbf{r}_{i+1}^{(FCF)}\|}{\|\mathbf{r}_i^{(FCF)}\|} < \varphi_{FCF}$$

for iterations $i > 1$ (i.e. not the first iteration).[1]

Broadly, the TAP defines how well k steps of the fine-grid time-stepping operator, Φ^k, must approximate the action of the coarse-grid operator, Ψ, for convergence.[2]

---

[1] The original derivations in [17] did not include the $-i$ in terms of N_c, but this is important to represent the exactness property of Parareal and two-level MGRIT, where convergence is exact after N_c iterations.

[2] There are some nuances regarding error versus residual and powers of operators/multiple iterations. We refer the interested reader to [17] for details.

An interesting property of the TAP is the introduction of a complex scaling of Ψ, even in the case of real-valued operators. If we suppose Ψ has imaginary eigenvalues, the minimum over x can be thought of as rotating this eigenvalue to the real axis. Recall from [17], the TAP for a fixed x_0 can be computed as the largest generalized singular value of $\left\{\Psi - \Phi^k, I - e^{\mathrm{i}x_0}\Psi\right\}$ or, equivalently, the largest singular value of $(\Psi - \Phi^k)(I - e^{\mathrm{i}x_0}\Psi)^{-1}$. In the numerical results in Sect. 3, we directly compute the generalized singular value decomposition (GSVD) of $\left\{\Psi - \Phi^k, I - e^{\mathrm{i}x_0}\Psi\right\}$ for a set of $x \in [0, 2\pi]$ to accurately evaluate the TAP, and refer to such as "GSVD bounds".

Although there are methods to compute the GSVD directly or iteratively, e.g. [21], minimizing the TAP for all $x \in [0, 2\pi]$ is expensive and often impractical. The following lemma and corollaries introduce a closed form for the minimum over x, and simplified sufficient conditions for convergence that do not require a minimization or complex operators.

Lemma 1 *Suppose Ψ is real-valued. Then,*

$$\min_{x\in[0,2\pi]} \|(I - e^{\mathrm{i}x}\Psi)\mathbf{v}\|^2 = \|\mathbf{v}\|^2 + \|\Psi\mathbf{v}\|^2 - 2\left|\langle\Psi\mathbf{v}, \mathbf{v}\rangle\right|, \tag{10}$$

$$\min_{x\in[0,2\pi]} \|\Phi^{-k}(I - e^{\mathrm{i}x}\Psi)\mathbf{v}\|^2 = \|\Phi^{-k}\mathbf{v}\|^2 + \|\Phi^{-k}\Psi\mathbf{v}\|^2 - 2\left|\langle\Phi^{-k}\Psi\mathbf{v}, \Phi^{-k}\mathbf{v}\rangle\right|.$$

Proof Consider satisfying the TAP for real-valued operators and complex vectors. Expanding in inner products, the TAP is given by

$$\min_{x\in[0,2\pi]} \|(I - e^{\mathrm{i}x}\Psi)\mathbf{v}\|^2 = \min_{x\in[0,2\pi]} \|\mathbf{v}\|^2 + \|\Psi\mathbf{v}\|^2 - e^{\mathrm{i}x}\langle\Psi\mathbf{v}, \mathbf{v}\rangle - e^{-\mathrm{i}x}\langle\mathbf{v}, \Psi\mathbf{v}\rangle.$$

Now decompose $\mathbf{v}$ into real and imaginary components, $\mathbf{v} = \mathbf{v}_r + \mathrm{i}\mathbf{v}_i$ for $\mathbf{v}_i, \mathbf{v}_r \in \mathbb{R}^n$, and note that

$$\begin{aligned}
\langle\Psi\mathbf{v}, \mathbf{v}\rangle &= \langle\Psi\mathbf{v}_i, \mathbf{v}_i\rangle + \langle\Psi\mathbf{v}_r, \mathbf{v}_r\rangle + \mathrm{i}\langle\Psi\mathbf{v}_i, \mathbf{v}_r\rangle - \mathrm{i}\langle\Psi\mathbf{v}_r, \mathbf{v}_i\rangle := \mathcal{R} - \mathrm{i}\mathcal{I},\\
\langle\mathbf{v}, \Psi\mathbf{v}\rangle &= \langle\mathbf{v}_i, \Psi\mathbf{v}_i\rangle + \langle\mathbf{v}_r, \Psi\mathbf{v}_r\rangle + \mathrm{i}\langle\mathbf{v}_i, \Psi\mathbf{v}_r\rangle - \mathrm{i}\langle\mathbf{v}_r, \Psi\mathbf{v}_i\rangle := \mathcal{R} + \mathrm{i}\mathcal{I}.
\end{aligned}$$

Expanding with $e^{\mathrm{i}x} = \cos(x) + \mathrm{i}\sin(x)$ and $e^{-\mathrm{i}x} = \cos(x) - \mathrm{i}\sin(x)$ yields

$$e^{\mathrm{i}x}\langle\Psi\mathbf{v}, \mathbf{v}\rangle + e^{-\mathrm{i}x}\langle\mathbf{v}, \Psi\mathbf{v}\rangle = 2\cos(x)\mathcal{R} + 2\sin(x)\mathcal{I}. \tag{11}$$

To minimize in x, differentiate and set the derivative equal to zero to obtain roots $\{x_0, x_1\}$, where $x_0 = \arctan(\mathcal{I}/\mathcal{R})$ and $x_1 = x_0 + \pi$. Plugging in above yields

$$\begin{aligned}
\min_{x\in[0,2\pi]} \pm\left(e^{\mathrm{i}x}\langle\Psi\mathbf{v}, \mathbf{v}\rangle + e^{-\mathrm{i}x}\langle\mathbf{v}, \Psi\mathbf{v}\rangle\right) &= -2\sqrt{\mathcal{R}^2 + \mathcal{I}^2}\\
&= -2\sqrt{(\langle\mathbf{v}_i, \Psi\mathbf{v}_i\rangle + \langle\mathbf{v}_r, \Psi\mathbf{v}_r\rangle)^2 + (\langle\Psi\mathbf{v}_i, \mathbf{v}_r\rangle - \langle\Psi\mathbf{v}_r, \mathbf{v}_i\rangle)^2}\\
&= -2\sqrt{(\langle\Psi\mathbf{v}, \mathbf{v}\rangle)^* \langle\Psi\mathbf{v}, \mathbf{v}\rangle}\\
&= -2\left|\langle\Psi\mathbf{v}, \mathbf{v}\rangle\right|.
\end{aligned}$$

Then,

$$\min_{x\in[0,2\pi]} \|(I - e^{\mathrm{i}x}\Psi)\mathbf{v}\|^2 = \min_{x\in[0,2\pi]} \|\mathbf{v}\|^2 + \|\Psi\mathbf{v}\|^2 - e^{\mathrm{i}x}\langle\Psi\mathbf{v}, \mathbf{v}\rangle - e^{-\mathrm{i}x}\langle\mathbf{v}, \Psi\mathbf{v}\rangle$$
$$= \|\mathbf{v}\|^2 + \|\Psi\mathbf{v}\|^2 - 2\,|\langle\Psi\mathbf{v}, \mathbf{v}\rangle|\,.$$

Analogous derivations hold for the FCF-TAP with a factor of Φ^{-k}.

Corollary 1 (Symmetry in x) *For real-valued operators, the TAP is symmetric in x when considered over all $\mathbf{v}$, that is, it is sufficient to consider $x \in [0, \pi]$.*

Proof From the above proof, suppose that $\mathbf{v} := \mathbf{v}_r + \mathrm{i}\mathbf{v}_i$ is minimized by $x_0 = \arctan(\mathcal{I}/\mathcal{R})$. Then, swap the sign on $\mathbf{v}_i \mapsto \hat{\mathbf{v}} := \mathbf{v}_r - \mathrm{i}\mathbf{v}_i$, which yields $\hat{\mathcal{I}} = -\mathcal{I}$ and $\hat{\mathcal{R}}=\mathcal{R}$, and $\hat{\mathbf{v}}$ is minimized at $\hat{x}_0 = \arctan(\hat{\mathcal{I}}/\hat{\mathcal{R}}) = \arctan(-\mathcal{I}/\mathcal{R}) = -\arctan(\mathcal{I}/\mathcal{R}) = -x_0$. Further note the equalities $\min_{x\in[0,2\pi]} \|(I - e^{\mathrm{i}x}\Psi)\mathbf{v}\| = \min_{x\in[0,2\pi]} \|(I - e^{\mathrm{i}x}\Psi)\hat{\mathbf{v}}\|$ and $\|(\Psi - \Phi^k)\mathbf{v}\| = \|(\Psi - \Phi^k)\hat{\mathbf{v}}\|$ and, thus, $\mathbf{v}$ and $\hat{\mathbf{v}}$ satisfy the F-TAP with the same constant, and x-values x_0 and $-x_0$. Similar derivations hold for the FCF-TAP.

Corollary 2 (A sufficient condition for the TAP) *For real-valued operators, sufficient conditions to satisfy the F-TAP and FCF-TAP, respectively, are that for all vectors $\mathbf{v}$,*

$$\|(\Psi - \Phi^k)\mathbf{v}\| \leq \varphi_F \cdot \mathrm{abs}(\|\mathbf{v}\| - \|\Psi\mathbf{v}\|), \tag{12}$$
$$\|(\Psi - \Phi^k)\Phi^k\mathbf{v}\| \leq \varphi_{FCF} \cdot \mathrm{abs}(\|\mathbf{v}\| - \|\Phi^{-k}\Psi\Phi^k\mathbf{v}\|). \tag{13}$$

Proof Note

$$\min_{x\in[0,2\pi]} \|(I - e^{\mathrm{i}x}\Psi)\mathbf{v}\|^2 \geq \|\mathbf{v}\|^2 + \|\Psi\mathbf{v}\|^2 - 2\|\Psi\mathbf{v}\|\,\|\mathbf{v}\|$$
$$= (\|\mathbf{v}\| - \|\Psi\mathbf{v}\|)^2\,.$$

Then, $\|(\Psi - \Phi^k)\mathbf{v}\| \leq \varphi_F|\|\Psi\mathbf{v}\| - \|\mathbf{v}\|| \leq \varphi_F \min_{x\in[0,2\pi]} \|(I - e^{\mathrm{i}x}\Psi)\mathbf{v}\|$. Similar derivations hold for $\min_{x\in[0,2\pi]} \|\Phi^{-k}(I - e^{\mathrm{i}x}\Psi)\mathbf{v}\|$.

For all practical purposes, a computable approximation to the TAP is sufficient, because the underlying purpose is to understand the convergence of Parareal and MGRIT and help pick or develop effective coarse-grid propagators. To that end, one can approximate the TAP by only considering it for specific x. For example, one could only consider real-valued $e^{\mathrm{i}x}$, with $x \in \{0, \pi\}$ (or, equivalently, only real-valued $\mathbf{v}$). Let $\Psi = \Psi_s + \Psi_k$, where $\Psi_s := (\Psi + \Psi^T)/2$ and $\Psi_k := (\Psi - \Psi^T)/2$ are the symmetric and skew-symmetric parts of Ψ. Then, from Lemma 1 and Remark 4 in [17], the TAP restricted to $x \in \{0, \pi\}$ takes the simplified form

$$\min_{x\in\{0,\pi\}} \|(I - e^{\mathrm{i}x}\Psi)\mathbf{v}\|^2 = \|\mathbf{v}\|^2 + \|\Psi\mathbf{v}\|^2 - 2\,|\langle\Psi_s\mathbf{v},\mathbf{v}\rangle|$$

$$= \begin{cases} \|(I+\Psi)\mathbf{v}\|^2 & \text{if } \langle\Psi\mathbf{v},\mathbf{v}\rangle \le 0 \\ \|(I-\Psi)\mathbf{v}\|^2 & \text{if } \langle\Psi\mathbf{v},\mathbf{v}\rangle > 0 \end{cases}. \tag{14}$$

Here, we have approximated the numerical range $|\langle\Psi\mathbf{v},\mathbf{v}\rangle|$ in (10) with the numerical range restricted to the real axis (i.e. the numerical range of the symmetric component of Ψ). If Ψ is symmetric, $\Psi_s = \Psi$, Ψ is unitarily diagonalizable, and the eigenvalue-based convergence bounds of [17] immediately pop out from (14). Because the numerical range is convex and symmetric across the real axis, (14) provides a reasonable approximation when Ψ has a significant symmetric component.

Now suppose Ψ is skew symmetric. Then $\Psi_s = \mathbf{0}$, and the numerical range lies exactly on the imaginary axis (corresponding with the eigenvalues of a skew symmetric operator). This suggests an alternative approximation to (10) by letting $x \in \{\pi/2, 3\pi/2\}$, which yields $e^{\mathrm{i}x} = \pm\mathrm{i}$. Similar to above, this yields a simplified version of the TAP,

$$\min_{x\in\{\pi/2,3\pi/2\}} \|(I - e^{\mathrm{i}x}\Psi)\mathbf{v}\|^2 = \|\mathbf{v}\|^2 + \|\Psi\mathbf{v}\|^2 - 2\,|\langle\Psi_k\mathbf{v},\mathbf{v}\rangle| \tag{15}$$

$$= \begin{cases} \|(I+\mathrm{i}\Psi)\mathbf{v}\|^2 & \text{if } \langle\Psi\mathbf{v},\mathbf{v}\rangle \le 0 \\ \|(I-\mathrm{i}\Psi)\mathbf{v}\|^2 & \text{if } \langle\Psi\mathbf{v},\mathbf{v}\rangle > 0 \end{cases}.$$

Recall skew-symmetric operators are normal and unitarily diagonalizable. Pulling out the eigenvectors in (15) and doing a maximum over eigenvalues again yields exactly the two-grid eigenvalue bounds derived in [17]. In particular, the $e^{\mathrm{i}x}$ is a rotation of the purely imaginary eigenvalues to the real axis, and corresponds to the denominator $1 - |\mu_i|$ of two-grid eigenvalue convergence bounds [17].

Finally, we point out that the simplest approximation of the TAP when Φ and Ψ share eigenvectors (notably, when they are based on the same spatial discretization) is to satisfy Definition 1 for all eigenvectors. We refer to this as the "Eval bound" in Sect. 3; note that the Eval bound is equivalent to the more general (and accurate) GSVD bound for normal operators when the eigenvectors form an orthogonal basis. For non-normal operators, Eval bounds are tight in the $(UU^*)^{-1}$-norm, for eigenvector matrix U. How well this represents convergence in the more standard ℓ^2-norm depends on the conditioning of the eigenvectors [17].

2.4 Two-Level Results and Why Multilevel Is Harder

Theorem 1 covers the case of two-level convergence in a fairly general (linear) setting. In practice, however, multilevel methods are often superior to two-level methods, so a natural question is what these results mean in the multilevel setting. Two-grid convergence usually provides a lower bound on possible multilevel convergence factors.

For MGRIT, it is known that F-cycles can be advantageous, or even necessary, for fast, scalable (multilevel) convergence [5], better approximating a two-level method on each level of the hierarchy than a normal V-cycle. However, because MGRIT uses non-Galerkin coarse grid operators, the relationship between two-level and multi-level convergence is complicated, and it is not obvious that two-grid convergence does indeed provide a lower bound on multilevel convergence.

Theorem 1 and results in [4, 17], can be interpreted in two ways. The first, and the interpretation used here, is that the derived tight bounds on the worst-case convergence factor hold for all but the first iteration. In [4], a different perspective is taken, where bounds in Theorem 1 hold for propagation of C-point error on all iterations.[3] In the two-level setting, either case provides necessary and sufficient conditions for convergence, assuming that the other iteration is bounded (which it is [17]). However, satisfying Definition 1 and Theorem 1 can still result in a method in which the ℓ^2-norm of the error of residual *grows* in the first iteration, when measured over all time points.

In the multilevel setting, we are indeed interested in convergence over *all* points for a *single* iteration. Consider a three-level MGRIT V-cycle, with levels 0, 1, and 2. On level 1, one iteration of two-level MGRIT is applied as an approximate residual correction for the problem on level 0. Suppose Theorem 1 ensures convergence for one iteration, that is, $\varphi < 1$. Because we are only performing one iteration, we must take the perspective that Theorem 1 ensures a decrease in C-point error, but a possible increase in F-point error on level 1. However, if the *total* error on level 1 has increased, coarse-grid correction on level 0 interpolates a *worse* approximation to the desired (exact) residual correction than interpolating no correction at all! If divergent behavior occurs somewhere in the middle of the multigrid hierarchy, it is likely that the larger multigrid iteration will also diverge. Given that multilevel convergence is usually worse than two level in practice, this motivates a stronger two-grid result that can ensure two-grid convergence for all points on all iterations.

The following theorem introduces stronger variations in the TAP (in particular, the conditions of Theorem 2 are stricter than, and imply, the TAP (Definition 1)), that provide necessary conditions for a two-level method with F- and FCF-relaxation to converge every iteration on C-points and F-points. Notationally, Theorem 1 provides necessary and sufficient conditions to bound the $\sim$-operators in (5), while here we provide necessary conditions to bound the full residual propagation operators, $\mathcal{R}_F$ and $\mathcal{R}_{FCF}$. Corollary 3 strengthens this result in the case of simultaneous diagonalization of Φ and Ψ, deriving a tight upper and lower bounds in norm of full error and residual propagation operators. For large N_c, having ensuring this bound < 1 provides necessary *and sufficient* conditions for guaranteed reduction in error and residual, over all points, in the first iteration.

Theorem 2 *Let* $\mathcal{R}_F$ *and* $\mathcal{R}_{FCF}$ *denote residual propagation of two-level MGRIT with F-relaxation and FCF-relaxation, respectively. Define* $\mathcal{W}_F := \sqrt{\sum_{\ell=0}^{k-1} \Phi^\ell (\Phi^\ell)^*}$,

---

[3] In [4] it is assumed that Φ and Ψ are unitarily diagonalizable, and upper bounds on convergence are derived. These bounds were generalized in [17] and shown to be tight in the unitarily diagonalizable setting.

$\mathcal{W}_{FCF} := \sqrt{\sum_{\ell=k}^{2k-1} \Phi^{\ell}(\Phi^{\ell})^*}$, *and* $\widehat{\varphi}_F$ *and* $\widehat{\varphi}_{FCF}$ *as the minimum constants such that, for all* $\mathbf{v}$,

$$\|(\Psi - \Phi^k)\mathbf{v}\| \leq \widehat{\varphi}_F \left[\min_{x\in[0,2\pi]} \left\| \mathcal{W}_F^{-1}(I - e^{ix}\Psi)\mathbf{v} \right\| + O(1/\sqrt{N_c}) \right],$$

$$\|(\Psi - \Phi^k)\mathbf{v}\| \leq \widehat{\varphi}_{FCF} \left[\min_{x\in[0,2\pi]} \left\| \mathcal{W}_{FCF}^{-1}(I - e^{ix}\Psi)\mathbf{v} \right\| + O(1/\sqrt{N_c}) \right].$$

Then, $\|\mathcal{R}_F\| \geq \widehat{\varphi}_F$ *and* $\|\mathcal{R}_{FCF}\| \geq \widehat{\varphi}_{FCF}$.

Proof The proof follows the theoretical framework developed in [17] and can be found in the appendix.

Corollary 3 *Assume that* Φ *and* Ψ *are simultaneously diagonalizable with eigenvectors* U *and eigenvalues* $\{\lambda_i\}_i$ *and* $\{\mu_i\}_i$*, respectively. Denote error and residual propagation operators of two-level MGRIT as* $\mathcal{E}$ *and* $\mathcal{R}$*, respectively, with subscripts indicating relaxation scheme. Let* $\widetilde{U}$ *denote a block-diagonal matrix with diagonal blocks given by* U*. Then,*

$$\begin{aligned} \|\mathcal{R}_F\|_{(\widetilde{U}\widetilde{U}^*)^{-1}} = \|\mathcal{E}_F\|_{(\widetilde{U}\widetilde{U}^*)^{-1}} &= \max_i \sqrt{\frac{1-|\lambda_i|^{2k}}{1-|\lambda_i|^2}} \frac{|\mu_i - \lambda_i^k|}{(1-|\mu_i|) + O(1/N_c)}, \\ \|\mathcal{R}_{FCF}\|_{(\widetilde{U}\widetilde{U}^*)^{-1}} = \|\mathcal{E}_{FCF}\|_{(\widetilde{U}\widetilde{U}^*)^{-1}} &= \max_i \sqrt{\frac{1-|\lambda_i|^{2k}}{1-|\lambda_i|^2}} \frac{|\lambda_i|^k|\mu_i - \lambda_i^k|}{(1-|\mu_i|) + O(1/N_c)}. \end{aligned} \tag{16}$$

Proof The proof follows the theoretical framework developed in [17] and can be found in the appendix.

For a detailed discussion on the simultaneously diagonalizable assumption, see [17].

Note in Theorem 2 and Corollary 3, there is an additional scaling compared with results obtained on two-grid convergence for all but the first iteration (see Theorem 1 and [17]), which makes the convergence bounds larger in all cases. This factor may be relevant to why multilevel convergence is more difficult than two-level. Figure 1 demonstrates the impact of this additional scaling by plotting the difference between worst-case two-grid convergence on all points on all iterations (Corollary 3) versus error on all points on all but one iteration (Theorem 1). Plots are a function of δt times the spatial eigenvalues in the complex plane. Note, the color map is strictly positive because worst-case error propagation on the first iteration is strictly greater than that on further iterations.

There are a few interesting points to note from Fig. 1. First, the second-order L-stable scheme yields good convergence over a far larger range of spatial eigenvalues and time steps than the A-stable scheme. The better performance of L-stable schemes is discussed in detail in [6], primarily for the case of SPD spatial operators. However, some of the results extend to the complex setting as well. In particular,

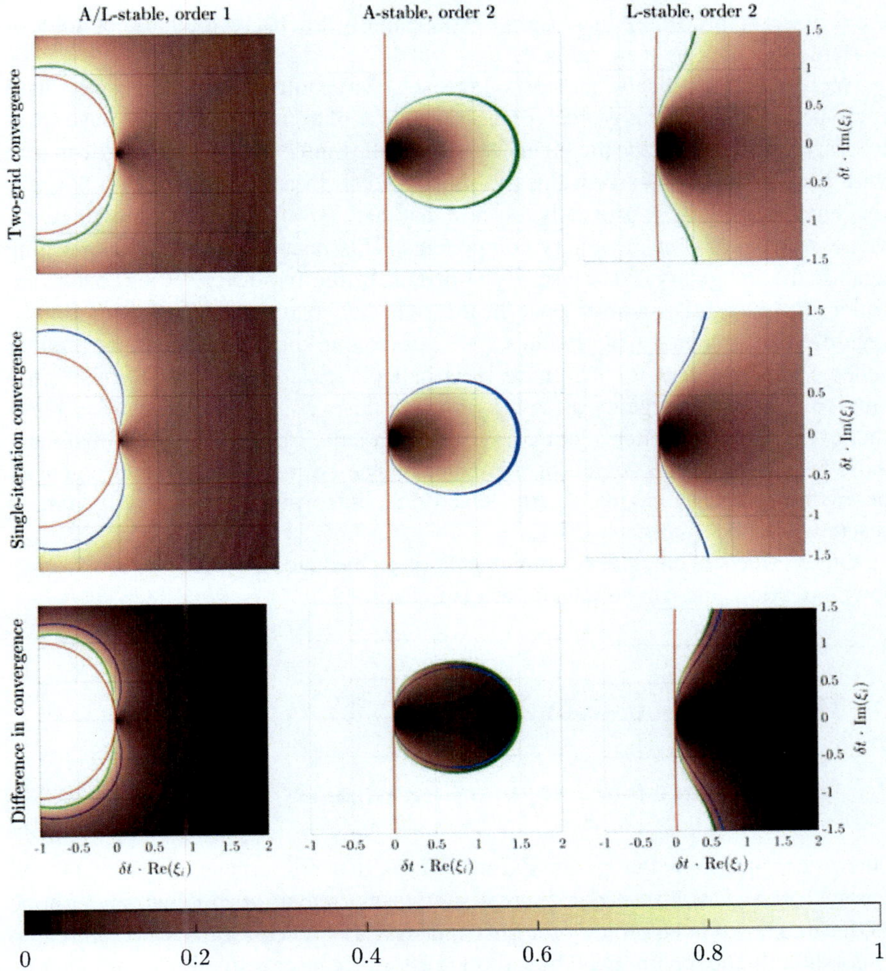

Fig. 1 Convergence bounds for two-level MGRIT with F-relaxation and $k = 4$, for A- and L-stable SDIRK schemes, of order 1 and 2, as a function of spatial eigenvalues $\{\xi_i\}$ in the complex plane. Red lines indicate the stability region of the integration scheme (stable left of the line). The top row shows two-grid convergence rates for all but one iteration (8), with the green line marking the boundary of convergence. Similarly, the second row shows single-iteration two-grid convergence (16), with blue line marking the boundary of convergence. The final row shows the difference in the convergence between single-iteration and further two-grid iterations

if Ψ is L-stable, then as $\delta t|\xi_i| \to \infty$ two-level MGRIT is convergent. This holds even on the imaginary axis, a spatial eigenvalue regime known to cause difficulties for parallel-in-time algorithms. As a consequence, it is possible there are compact regions in the positive half plane where two-level MGRIT will not converge, but convergence is guaranteed at the origin and outside of these isolated regions. Convergence for large time steps is particularly relevant for multilevel schemes because the coarsening procedure increases δt exponentially. Such a result does not hold for A-stable schemes, as seen in Fig. 1.

Second, Fig. 1 indicates (at least one reason) why multilevel convergence is hard for hyperbolic PDEs. It is typical for discretizations of hyperbolic PDEs to have spectra that push up against the imaginary axis. From the limit of a skew-symmetric matrix with purely imaginary eigenvalues to more diffusive discretizations with nonzero real eigenvalue parts, it is usually the case that there are small (in magnitude) eigenvalues with dominant imaginary components. This results in eigenvalues pushing against the imaginary axis close to the origin. In the two-level setting, backward Euler still guarantees convergence in the right half plane (see top left of Fig. 1), regardless of imaginary eigenvalues. Note that to our knowledge, no other implicit scheme is convergent for the entire right half plane. However, when considering single-iteration convergence as a proxy for multilevel convergence, we see in Fig. 1 that even backward Euler is not convergent in a small region along the imaginary axis. In particular, this region of non-convergence corresponds to small eigenvalues with dominant imaginary parts, exactly the eigenvalues that typically arise in discretizations of hyperbolic PDEs.

Other effects of imaginary components of eigenvalues are discussed in the following section, and can be seen in results in Sect. 3.3.

3 Theoretical Bounds in Practice

3.1 *Convergence and Imaginary Eigenvalues*

One important point that follows from [17] is that convergence of MGRIT and Parareal *only depends on the discrete spatial and temporal problem.* Hyperbolic PDEs are known to be difficult for PinT methods. However, for MGRIT/Parareal, it is not directly the (continuous) hyperbolic PDE that causes difficulties, but rather its discrete representation. Spatial discretizations of hyperbolic PDEs (when diagonalizable) often have eigenvalues with relatively large imaginary components compared to the real component, particularly as magnitude $|\lambda| \to 0$. In this section, we look at why eigenvalues with dominant imaginary part are difficult for MGRIT and Parareal. The results are limited to diagonalizable operators, but give new insight into (the known fact) that diffusivity of the backward Euler scheme makes it more amenable to PinT methods. We also look at the relation of temporal problem size and coarsening factor to convergence, which is particularly important for such discretizations, as well as the disappointing acceleration of FCF-relaxation. Note, least-squares discretizations have been developed for hyperbolic PDEs that result in an SPD spatial matrix (e.g. [12]), which would immediately overcome the problems that arise with complex/imaginary eigenvalues discussed in this section. Whether a given discretization provides the desired approximation to the continuous PDE is a different question.

Problem size and FCF-relaxation: Consider the exact solution to the linear time propagation problem, $\mathbf{u}_t = \mathcal{L}\mathbf{u}$, given by

$$\hat{\mathbf{u}} := e^{-\mathcal{L}t}\mathbf{u}.$$

Then, an exact time step of size δt is given by $\mathbf{u} \leftarrow e^{-\mathcal{L}\delta t}\mathbf{u}$. Runge-Kutta schemes are designed to approximate this (matrix) exponential as a rational function, to order δt^p for some p. Now suppose $\mathcal{L}$ is diagonalizable. Then, propagating the ith eigenvector, say $\mathbf{v}_i$, forward in time by δt, is achieved through the action $e^{-\delta t \xi_i}\mathbf{v}_i$, where ξ_i is the ith corresponding eigenvalue. Then the exact solution to propagating $\mathbf{v}_i$ forward in time by δt is given by

$$e^{-\delta t \xi_i} = e^{-\delta t \mathrm{Re}(\xi_i)}\Big(\cos(\delta t \mathrm{Im}(\xi_i)) - \mathrm{i}\sin(\delta t \mathrm{Im}(\xi_i))\Big). \tag{17}$$

If ξ_i is purely imaginary, raising (17) to a power k, corresponding to k time steps, yields the function $e^{\pm \mathrm{i}k\delta t|\xi_i|} = \cos(k\delta t|\xi_i|) \pm \mathrm{i}\sin(k\delta t|\xi_i|)$. This function is (i) magnitude one for all k, δt, and ξ, and (ii) performs exactly k revolutions around the origin in a circle of radius one. Even though we do not compute the exact exponential when integrating in time, we do approximate it, and this perspective gives some insight into why operators with imaginary eigenvalues tend to be particularly difficult for MGRIT and Parareal.

Recall the convergence bounds developed and stated in Sect. 2 as well as [17] have a $\mathcal{O}(1/\sqrt{N_c})$ term in the denominator. In many cases, the $\mathcal{O}(1/\sqrt{N_c})$ term is in some sense arbitrary and the bounds in Theorem 1 are relatively tight for practical N_c. However, for some problems with spectra or field-of-values aligned along the imaginary axis, convergence can differ significantly as a function of N_c (see [6]). To that end, we restate Theorem 30 from [17], which provides tight upper and lower bounds, including the constants on N_c, for the case of diagonalizable operators.

Theorem 3 (Tight bounds—the diagonalizable case) *Let Φ denote the fine-grid time-stepping operator and Ψ denote the coarse-grid time-stepping operator, with coarsening factor k, and N_c coarse-grid time points. Assume that Φ and Ψ commute and are diagonalizable, with eigenvectors as columns of U, and spectra $\{\lambda_i\}$ and $\{\mu_i\}$, respectively. Then, worst-case convergence of error (and residual) in the $(UU^*)^{-1}$-norm is exactly bounded by*

$$\sup_j \left(\frac{|\mu_j - \lambda_j^k|}{\sqrt{(1-|\mu_j|)^2 + \frac{\pi^2|\mu_j|}{N_c^2}}} \right) \le \frac{\|\mathbf{e}_{i+1}^{(F)}\|_{(UU^*)^{-1}}}{\|\mathbf{e}_i^{(F)}\|_{(UU^*)^{-1}}} \le \left(\sup_j \frac{|\mu_j - \lambda_j^k|}{\sqrt{(1-|\mu_j|)^2 + \frac{\pi^2|\mu_j|}{6N_c^2}}} \right),$$

$$\sup_j \left(\frac{|\lambda_j^k||\mu_j - \lambda_j^k|}{\sqrt{(1-|\mu_j|)^2 + \frac{\pi^2|\mu_j|}{N_c^2}}} \right) \le \frac{\|\mathbf{e}_{i+1}^{(FCF)}\|_{(UU^*)^{-1}}}{\|\mathbf{e}_i^{(FCF)}\|_{(UU^*)^{-1}}} \le \left(\sup_j \frac{|\lambda_j^k||\mu_j - \lambda_j^k|}{\sqrt{(1-|\mu_j|)^2 + \frac{\pi^2|\mu_j|}{6N_c^2}}} \right),$$

for all but the last iteration (or all but the first iteration for residual).

The counter-intuitive nature of MGRIT and Parareal convergence is that convergence is defined by how well k steps on the fine-grid *approximate the coarse-grid operator*. That is, in the case of eigenvalue bounds, we must have $|\mu_i - \lambda_i^k|^2 \lessapprox [(1-|\mu_i|)^2 + 10/N_c^2$, for each coarse-grid eigenvalue μ_i. Clearly, the important cases for convergence are $|\mu_i| \approx 1$. Returning to (17), purely imaginary spatial eigenvalues typically lead to $|\mu_i|, |\lambda_j| \approx 1$, particularly for small $|\delta t \xi_i|$ (because the RK approximation to the exponential is most accurate near the origin). This has several effects on convergence:

1. Convergence will deteriorate as the number of time steps increases,
2. λ_i^k must approximate μ_i highly accurately for many eigenvalues,
3. FCF-relaxation will offer little to no improvement in convergence, and
4. Convergence will be increasingly difficult for larger coarsening factors.

For the first and second points, notice from bounds in Theorem 3 that the order of $N_c \gg 1$ is generally only significant when $|\mu_i| \approx 1$. With imaginary eigenvalues, however, this leads to a moderate part of the spectrum in which λ_i^k must approximate μ_i with accuracy $\approx 1/N_c$, which is increasingly difficult as $N_c \to \infty$. Conversely, introducing real parts to spatial eigenvalues introduces dissipation in (17), particularly when raising to powers. Typically the result of this is that $|\mu_i|$ decreases, leading to fewer coarse-grid eigenvalues that need to be approximated well, and a lack of dependence on N_c.

In a similar vein, the third point above follows because in terms of convergence, FCF-relaxation adds a power of $|\lambda_i|^k$ to convergence bounds compared with F-relaxation. Improved convergence (hopefully due to FCF-relaxation) is needed when $|\mu_i| \approx 1$. In an unfortunate cyclical fashion, however, for such eigenvalues, it must be the case that $\lambda_i^k \approx \mu_i$. But if $|\mu_i| \approx 1$, and $\lambda_i^k \approx \mu_i$, the additional factor lent by FCF-relaxation, $|\lambda_i| \approx 1$, which tends to offer at best marginal improvements in convergence. Finally, point four, which specifies that it will be difficult to observe nice convergence with larger coarsening factors, is a consequence of points two and three. As k increases, Ψ must approximate a rational function, Φ^k, of polynomials with a progressively higher degree. When Ψ must approximate Φ well for many eigenmodes and, in addition, FCF-relaxation offers minimal improvement in convergence, a convergent method quickly becomes intractable.

Convergence in the complex plane: Although eigenvalues do not always provide a good measure of convergence (for example, see Sect. 3.3), they can provide invaluable information on choosing a good time integration scheme for Parareal/MGRIT. Some of the properties of a "good" time integration scheme transfer from the analysis of time integration to parallel-in-time, however, some integration schemes are far superior for parallel-in-time integration, without an obvious/intuitive reason why. This section demonstrates how we can analyze time-stepping schemes by considering the convergence of two-level MGRIT/Parareal as a function of eigenvalues in the complex plane. Figures 2 and 3 plot the real and imaginary parts of eigenvalues $\lambda \in \sigma(\Phi)$ and $\mu \in \sigma(\Psi)$ as a function of δt times the spatial eigenvalue, as well as the corresponding two-level convergence for F- and FCF-relaxation, for an A-stable

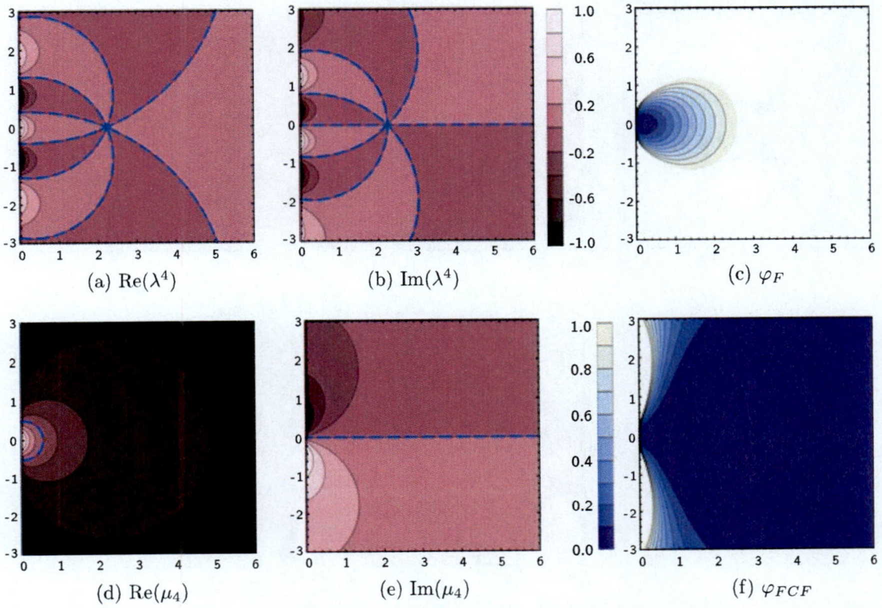

Fig. 2 Eigenvalues and convergence bounds for ESDIRK-33, $p = 3$ and $k = 4$. Dashed blue lines indicate the sign changes. Note, if we zoom out on the FCF-plot, it actually resembles that of F with a diameter of about 100 rather than 2. Thus, for $\delta t \xi \gg 1$, even FCF-relaxation does not converge well

ESDIRK-33 Runge-Kutta scheme and a third-order explicit Runge-Kutta scheme, respectively.

There are a few interesting things to note. First, FCF-relaxation expands the region of convergence in the complex plane dramatically for ESDIRK-33. However, there are other schemes (not shown here, for brevity) where FCF-relaxation provides little to no improvement. Also, note that the fine eigenvalue λ^4 changes sign many times along the imaginary axis (in fact, the real part of λ^k changes signs $2k$ times and the imaginary part $2k - 1$). Such behavior is very difficult to approximate with a coarse-grid time-stepping scheme and provides another way to think about why imaginary eigenvalues and hyperbolic PDEs can be difficult for Parareal MGRIT. On a related note, using explicit time-stepping schemes in Parareal/MGRIT is inherently limited by ensuring a stable time step on the coarse grid, which makes naive application rare in numerical PDEs. However, when stability is satisfied on the coarse grid, Fig. 3 (and similar plots for other explicit schemes) suggests that the domain of convergence pushes much closer against the imaginary axis for explicit schemes than implicit. Such observations may be useful in applying Parareal and MGRIT methods to systems of ODEs with imaginary eigenvalues, but less stringent stability requirements, wherein explicit integration may be a better choice than implicit.

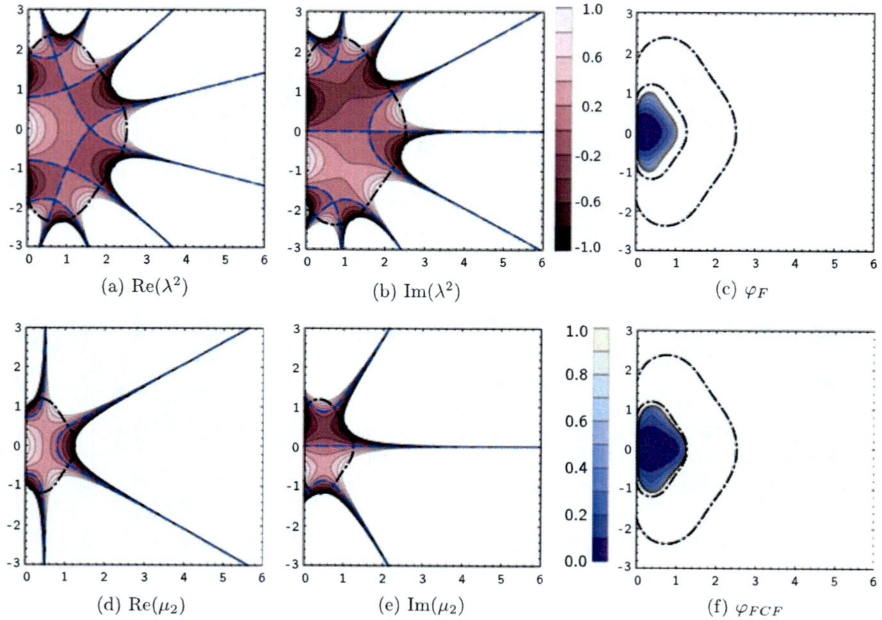

(a) $\mathrm{Re}(\lambda^2)$ (b) $\mathrm{Im}(\lambda^2)$ (c) φ_F

(d) $\mathrm{Re}(\mu_2)$ (e) $\mathrm{Im}(\mu_2)$ (f) φ_{FCF}

Fig. 3 Eigenvalues and convergence bounds for ERK, $p = 3$ and $k = 2$. Dashed blue lines the indicate sign changes

3.2 *Test Case: The Wave Equation*

The previous section considered the effects of imaginary eigenvalues on the convergence of MGRIT and Parareal. Although true skew-symmetric operators are not particularly common, a similar character can be observed in other discretizations. In particular, writing the second-order wave equation in first-order form and discretizing often leads to a spatial operator that is nearly skew-symmetric. Two-level and multilevel convergence theory for MGRIT based on eigenvalues was demonstrated to provide moderately accurate convergence estimates for small-scale discretizations of the second-order wave equation in [9]. Here, we investigate the second-order wave equation further in the context of a finer spatiotemporal discretization, examining why eigenvalues provide reliable information on convergence, looking at the single-iteration bounds from Corollary 3, and discussing the broader implications.

The second-order wave equation in two spatial dimensions over domain $\Omega = (0, 2\pi) \times (0, 2\pi)$ is given by $\partial_{tt}u = c^2 \Delta u$ for $\mathbf{x} \in \Omega$, $t \in (0, T]$ with scalar solution $u(\mathbf{x}, t)$ and wave speed $c = \sqrt{10}$. This can be written equivalently as a system of PDEs that are first order in time,

$$\begin{bmatrix} u \\ v \end{bmatrix}_t - \begin{bmatrix} 0 & I \\ c^2\Delta & 0 \end{bmatrix} \begin{bmatrix} u \\ v \end{bmatrix} = \begin{bmatrix} 0 \\ 0 \end{bmatrix}, \qquad \text{for } \mathbf{x} \in \Omega, t \in (0, T], \tag{18}$$

with initial and boundary conditions

$$\begin{aligned} &u(\cdot, 0) = \sin(x)\sin(y), \quad v(\cdot, 0) = 0, && \text{for } \mathbf{x} \in \Omega \cup \partial\Omega, \\ &u(\mathbf{x}, \cdot) = v(\mathbf{x}, \cdot) = 0, && \text{for } \mathbf{x} \in \partial\Omega. \end{aligned}$$

Why eigenvalues: Although the operator that arises from writing the second-order wave equation as a first-order system in time is not skew-adjoint, one can show that it (usually) has purely imaginary eigenvalues. Moreover, although not unitarily diagonalizable (in which case the $(UU^*)^{-1}$-norm equals the ℓ^2-norm; see Sect. 2.3), the eigenvectors are only ill-conditioned (i.e. not orthogonal) in a specific, localized sense. As a result, the Eval bounds provide an accurate measure of convergence, and a very good approximation to the formally ℓ^2-accurate (and much more difficult to compute) GSVD bounds.

Let $\{\mathbf{w}_\ell, \zeta_\ell\}$ be an eigenpair of the discretization of $-c^2\Delta u = 0$ used in (18), for $\ell = 0, ..., n-1$. For most standard discretizations, we have $\zeta_\ell > 0\ \forall \ell$ and the set $\{\mathbf{w}_\ell\}$ forms an orthonormal basis of eigenvectors. Suppose this is the case. Expanding the block eigenvalue problem $\mathbf{A}\mathbf{u}_j = \xi_j \mathbf{u}_j$ corresponding to (18),

$$\begin{bmatrix} \mathbf{0} & I \\ c^2\Delta & \mathbf{0} \end{bmatrix} \begin{bmatrix} \mathbf{x}_j \\ \mathbf{v}_j \end{bmatrix} = \xi_j \begin{bmatrix} \mathbf{x}_j \\ \mathbf{v}_j \end{bmatrix},$$

yields a set of $2n$ eigenpairs, grouped in conjugate pairs of the corresponding purely imaginary eigenvalues,

$$\begin{aligned} \{\mathbf{u}_{2\ell}, \xi_{2\ell}\} &:= \left\{ \frac{1}{\sqrt{1+\zeta_\ell}} \begin{bmatrix} \mathbf{w}_\ell \\ \mathrm{i}\sqrt{\zeta_\ell}\mathbf{w}_\ell \end{bmatrix}, \mathrm{i}\sqrt{\zeta_\ell} \right\}, \\ \{\mathbf{u}_{2\ell+1}, \xi_{2\ell+1}\} &:= \left\{ \frac{1}{\sqrt{1+\zeta_\ell}} \begin{bmatrix} \mathbf{w}_\ell \\ -\mathrm{i}\sqrt{\zeta_\ell}\mathbf{w}_\ell \end{bmatrix}, -\mathrm{i}\sqrt{\zeta_\ell} \right\}, \end{aligned}$$

for $\ell = 0, ..., n-1$. Although the $(UU^*)^{-1}$-norm can be expressed in closed form, it is rather complicated. Instead, we claim that eigenvalue bounds (theoretically tight in the $(UU^*)^{-1}$-norm) provide a good estimate of ℓ^2-convergence by considering the conditioning of eigenvectors.

Let U denote a matrix with columns given by eigenvectors $\{\mathbf{u}_j\}$, ordered as above for $\ell = 0, ..., n-1$. We can consider the conditioning of eigenvectors through the product

$$U^*U = \begin{bmatrix} 1 & \frac{1-\zeta_0}{1+\zeta_0} & & & \\ \frac{1-\zeta_0}{1+\zeta_0} & 1 & & & \\ & & \ddots & & \\ & & & 1 & \frac{1-\zeta_{n-1}}{1+\zeta_{n-1}} \\ & & & \frac{1-\zeta_{n-1}}{1+\zeta_{n-1}} & 1 \end{bmatrix}. \tag{19}$$

Notice that U^*U is a block-diagonal matrix with 2×2 blocks corresponding to conjugate pairs of eigenvalues. The eigenvalues of the 2×2 block are given by $\{2\zeta_\ell/(1+\zeta_\ell), 2/(1+\zeta_\ell)\}$. Although (19) can be ill-conditioned for large $\zeta_\ell \sim 1/h^2$, for spatial mesh size h, the ill-conditioning is only between conjugate pairs of eigenvalues, and eigenvectors are otherwise orthogonal. Furthermore, the following proposition proves that convergence bounds are symmetric across the real axis, that is, the convergence bound for eigenvector with spatial eigenvalue ξ is equivalent to that for its conjugate $\bar{\xi}$. Together, these facts suggest the ill-conditioning between eigenvectors of conjugate pairs will not significantly affect the accuracy of bounds and that tight eigenvalue convergence bounds for MGRIT in the $(UU^*)^{-1}$-norm provide accurate estimates of performance in practice.

Proposition 1 *Let Φ and Ψ correspond to Runge-Kutta discretizations in time, as a function of a diagonalizable spatial operator $\mathcal{L}$, with eigenvalues $\{\xi_i\}$. Then (tight) two-level convergence bounds of MGRIT derived in [17] as a function $\delta t\xi$ are symmetric across the real axis.*

Proof Recall from [6, 17], eigenvalues of Φ and Ψ correspond to the Runge-Kutta stability functions applied to $\delta t\xi$ and $k\delta t\xi$, respectively, for coarsening factor k. Also note that the stability function of a Runge-Kutta method can be written as a rational function of two polynomials with real coefficients, $P(z)/Q(z)$ [2]. As a result of the fundamental theorem of linear algebra $P(\bar{z})/Q(\bar{z}) = \overline{P(z)/Q(z)}$. Thus, for spatial eigenvalue ξ, $|\lambda(\xi)| = |\lambda(\bar{\xi})|$, $|\mu(\xi)| = |\mu(\bar{\xi})|$, and $|\mu(\bar{\xi}) - \lambda(\bar{\xi})^k| = |\overline{\mu(\xi) - \lambda(\xi)^k}| = |\mu(\xi) - \lambda(\xi)^k|$, which implies that convergence bounds in Theorem 3 are symmetric across the real axis.

Observed convergence versus bounds: The first-order form (18) is implemented in MPI/C++ using second-order finite differences in space and various L-stable SDIRK time integration schemes (see [9, Sect. SM3]). We consider 4096 points in the temporal domain and 41 points in the spatial domain, with $4096\delta t = 40\delta_x = T = 2\pi$ and $4096\delta t = 10 \cdot 40\delta_x = T = 20\pi$, such that $\delta t \approx 0.1\delta_x/c^2$ and $\delta t \approx \delta_x/c^2$, respectively. Two-level MGRIT with FCF-relaxation is tested for temporal coarsening factors $k \in \{2, 4, 8, 16, 32\}$ using the XBraid library [20], with a random initial space-time guess, an absolute convergence tolerance of 10^{-10}, and a maximum of 200 and 1000 iterations for SDIRK and ERK schemes, respectively. Figure 4 reports the geometric average ("Ave CF") and worst-case ("Worst CF") convergence factors for full MGRIT solves, along with estimates for the "Eval single it" bound from Corollary 3 by letting $N_c \to \infty$, the "Eval bound" as the eigenvalue approximation of the TAP (see Sect. 2.3), and the upper/lower bound from Theorem 3.

It turns out that for this particular problem and discretization, the convergence of Parareal/MGRIT requires a roughly explicit restriction on time step size ($\delta t < \delta x/c^2$, with no convergence guarantees at $\delta t = \delta x/c^2$). To that end, it is not likely to be competitive in practice versus sequential time stepping. Nevertheless, our eigenvalue-based convergence theory provides a very accurate measure of convergence. In all cases, the theoretical lower and upper bounds from Theorem 3 perfectly contain the

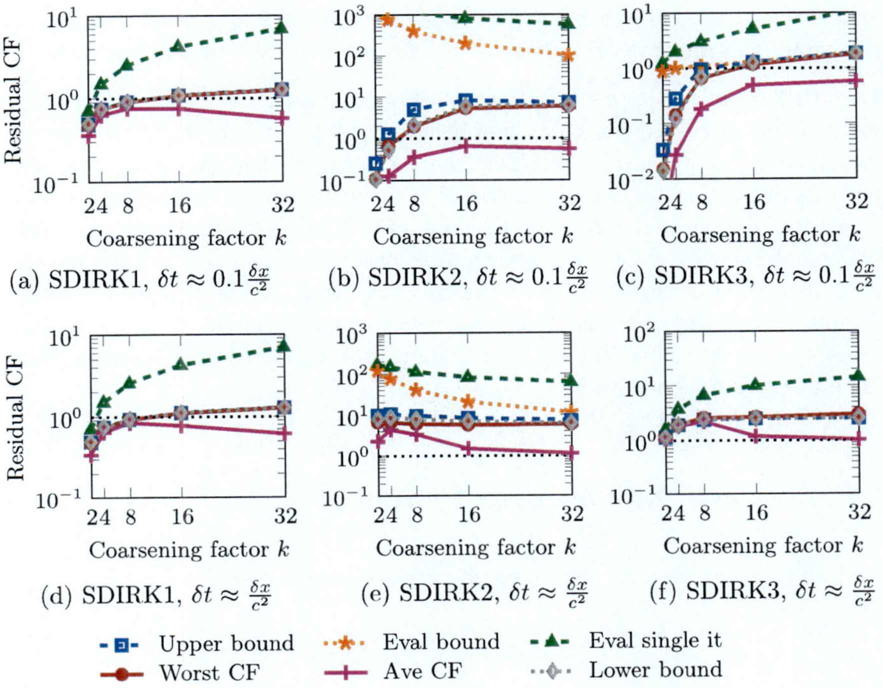

Fig. 4 Eigenvalue convergence bounds and upper/lower bounds from [17, Eq. (63)] compared to observed worst-case and average convergence factors

worst observed convergence factor, with only a very small difference between the theoretical upper and lower bounds.

There are a number of other interesting things to note from Fig. 4. First, both the general eigenvalue bound ("Eval bound") and single-iteration bound from Corollary 3 ("Eval single it") are very pessimistic, sometimes many orders of magnitude larger than the worst-observed convergence factor. As discussed previously and first demonstrated in [6], the size of the coarse problem, N_c, is particularly relevant for problems with imaginary eigenvalues. Although the lower and upper bound convergences to the "Eval bound" as $N_c \to \infty$, for small to moderate N_c, reasonable convergence may be feasible in practice even if the limiting bound $\gg 1$. Due to the relevance of N_c, it is difficult to comment on the single-iteration bound (because as derived, we let $N_c \to \infty$ for an upper single-iteration bound). However, we note that the upper bound on all other iterations ("Upper bound") appears to bound the worst-observed convergence factor, so the single-iteration bound on worst-case convergence appears not to be realized in practice, at least for this problem. It is also worth pointing out the difference between time integration schemes in terms of convergence, with backward Euler being the most robust, followed by SDIRK3, and last SDIRK2. As discussed in [6] for normal spatial operators, not all time-integration schemes are equal when it comes to convergence of Parareal and MGRIT, and robust convergence theory provides important information on choosing effective integration schemes.

Remark 1 (*Superlinear convergence*) In [8], superlinear convergence of Parareal is observed and discussed. We see similar behavior in Fig. 4, where the worst observed convergence factor can be more than ten times larger than the average convergence factor. In general, superlinear convergence would be expected as the exactness property of Parareal/MGRIT is approached. In addition, for non-normal operators, although the slowest converging mode may be observed in practice during some iteration, it does not necessarily stay in the error spectrum (and thus continue to yield slow(est) convergence) as iterations progress, due to the inherent rotation in powers of non-normal operators. The nonlinear setting introduces additional complications that may explain such behavior.

3.3 Test Case: DG Advection (Diffusion)

Now we consider the time-dependent advection-diffusion equation,

$$\frac{\partial u}{\partial t} + \mathbf{v} \cdot \nabla u - \nabla \cdot (\alpha \nabla u) = f, \tag{20}$$

on a 2D unit square domain discretized with linear discontinuous Galerkin elements. The discrete spatial operator, L, for this problem is defined by $L = M^{-1}K$, where M is a mass matrix, and K is the stiffness matrix associated with $\mathbf{v} \cdot \nabla u - \nabla \cdot (\alpha \nabla u)$. The length of the time domain is always chosen to maintain the appropriate relationship between the spatial and temporal step sizes while also keeping 100 time points on the MGRIT coarse grid (for a variety of coarsening factors). Throughout, the initial condition is $u(\mathbf{x}, 0) = \mathbf{0}$ and the following time-dependent forcing function is used:

$$f(\mathbf{x}, t) = \begin{cases} \cos^2(2\pi t / t_{final}), & \mathbf{x} \in [1/8, 3/8] \times [1/8, 3/8] \\ 0, & \text{else.} \end{cases} \tag{21}$$

Backward Euler and a third-order, three-stage L-stable SDIRK scheme, denoted SDIRK3 are applied in time, and FCF-relaxation is used for all tests. Three different velocity fields are studied:

$$\mathbf{v}_1(\mathbf{x}, t) = (\sqrt{2/3}, \sqrt{1/3}), \tag{22}$$

$$\mathbf{v}_2(\mathbf{x}, t) = (\cos(\pi y)^2, \cos(\pi x)^2), \tag{23}$$

$$\mathbf{v}_3(\mathbf{x}, t) = (y\pi/2, -x\pi/2), \tag{24}$$

referred to as problems 1, 2, and 3 below. Note that problem 1 is a simple translation, problem 2 is a curved but non-recirculating flow, and problem 3 is a recirculating flow.

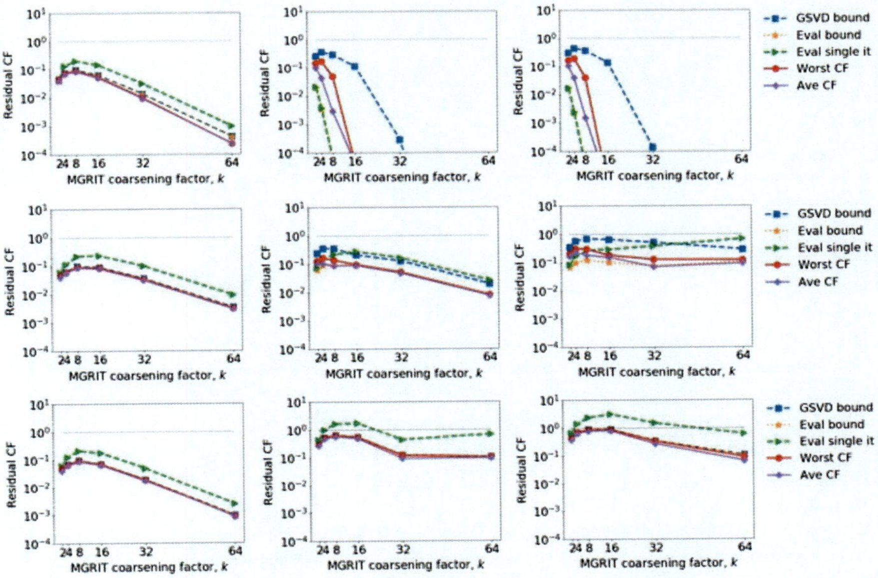

Fig. 5 Bounds and observed convergence factor versus MGRIT coarsening factor with $n = 16$ using backward Euler for problem 1 (top), 2 (middle), and 3 (bottom) with $\alpha = 10\delta x$ (left), $0.1\delta x$ (middle), and 0 (right)

The relative strength of the diffusion term is also varied in the results below, including a diffusion-dominated case, $\alpha = 10\delta x$, an advection-dominated case, $\alpha = 0.1\delta x$, and a pure advection case, $\alpha = 0$. When backward Euler is used, the time step is set equal to the spatial step, $\delta t = \delta x$, while for SDIRK3, $\delta t^3 = \delta x$, in order to obtain similar accuracy in both time and space.

Figure 5 shows various computed bounds compared with observed worst-case and average convergence factors (over ten iterations) versus MGRIT coarsening factor for each problem variation with different amounts of diffusion using backward Euler for time integration. The bounds shown are the "GSVD bound" from Theorem 1, the "Eval bound", an equivalent eigenvalue form of this bound (see [17, Theorem 13], and "Eval single it", which is the bound from Corollary 3 as $N_c \to \infty$. The problem size is $n = 16$, where the spatial mesh has $n \times n$ elements. This small problem size allows the bounds to be directly calculated: for the GSVD bound, it is possible to compute $||(\Psi - \Phi^k)(I - e^{\mathrm{i}x}\Psi)^{-1}||$, and for the eigenvalue bounds, it is possible to compute the complete eigenvalue decomposition of the spatial operator, L, and apply proper transformations to the eigenvalues to obtain eigenvalues of Φ and Ψ.

In the diffusion-dominated case (left column of Fig. 5), the GSVD and eigenvalue bounds agree well with each other (because the spatial operator is nearly symmetric), accurately predicting observed residual convergence factors for all problems. Similar to Sect. 3.2, the single-iteration bound from Corollary 3 does not appear to be realized in practice.

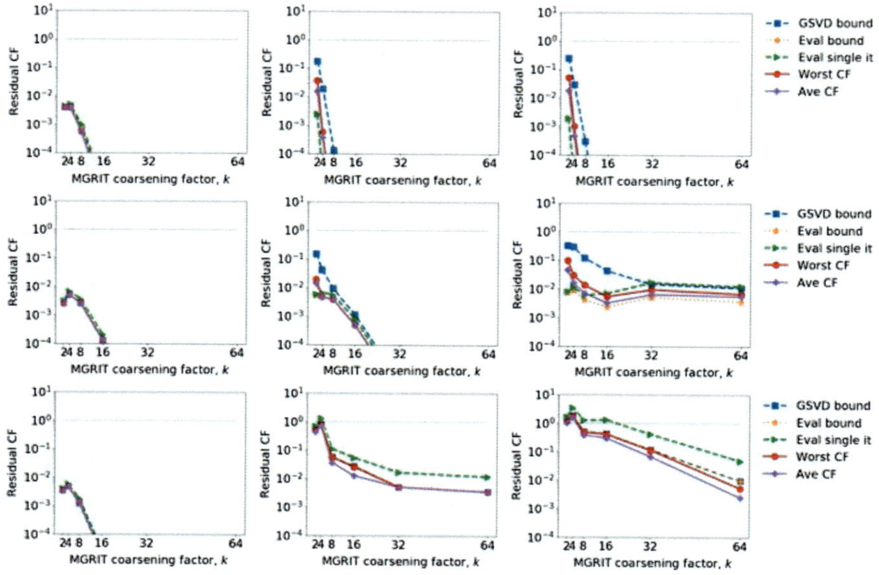

Fig. 6 Bounds and observed convergence factor versus MGRIT coarsening factor with $n = 16$ using SDIRK3 with $\delta t^3 = \delta x$ for problem 1 (top), 2 (middle), and 3 (bottom) with $\alpha = 10\delta x$ (left), $0.1\delta x$ (middle), and 0 (right)

In the advection-dominated and pure advection cases (middle and right columns of Fig. 5), behavior of the bounds and observed convergence depends on the type of flow. In the non-recirculating examples, the GSVD bounds are more pessimistic compared to observed convergence, but still provide an upper bound on worst-case convergence, as expected. Conversely, the eigenvalue bounds on worst-case convergence become unreliable, sometimes undershooting the observed convergence factors by significant amounts. Recall that the eigenvalue bounds are tight in the $(UU^*)^{-1}$-norm of the error, where U is the matrix of eigenvectors. However, for the non-recirculating problems, the spatial operator L is defective to machine precision, that is, the eigenvectors are nearly linearly dependent and U is close to singular. Thus, tight bounds on convergence in the $(UU^*)^{-1}$-norm are unlikely to provide an accurate measure of convergence in more standard norms, such as ℓ^2. In the recirculating case, UU^* is well conditioned. Then, similarly to the wave equation in Sect. 3.2, the eigenvalue bounds should provide a good approximation to the ℓ^2-norm, and, indeed, the GSVD and eigenvalue bounds agree well with each other accurately predict residual convergence factors.

Figure 6 shows the same set of results but using SDIRK3 for time integration instead of backward Euler and with a large time step $\delta t^3 = \delta x$ to match accuracy in the temporal and spatial domains. Results here are qualitatively similar to those of backward Euler, although MGRIT convergence (both predicted and measured) is generally much swifter, especially for larger coarsening factors. Again, the GSVD and eigenvalue bounds accurately predict observed convergence in the diffusion-

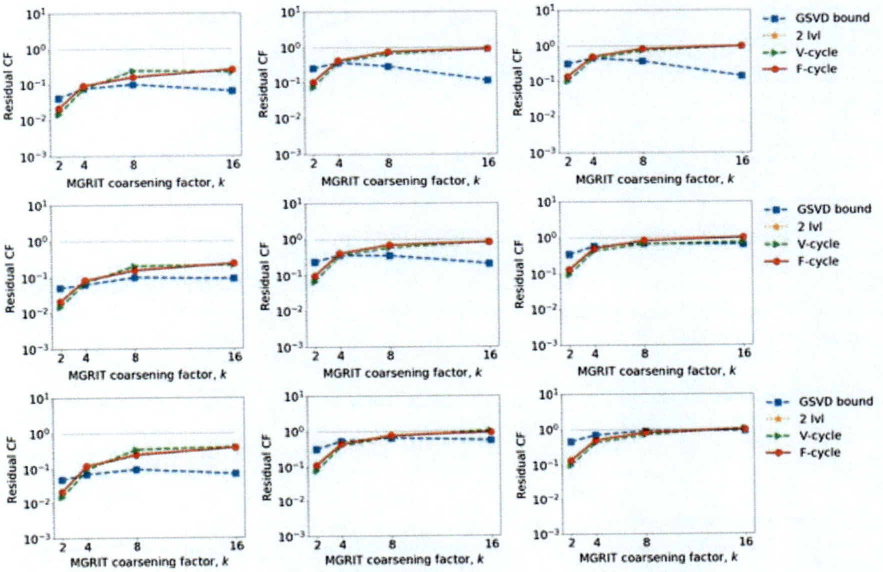

Fig. 7 GSVD bound for $n = 16$ versus observed convergence factors for different cycle structures with $n = 512$ plotted against MGRIT coarsening factor using backward Euler for problem 1 (top), 2 (middle), and 3 (bottom) with $\alpha = 1$ (left), 0.01 (middle), and 0.0 (right)

dominated case. In the advection-dominated and pure advection cases, again the eigenvalue bounds are not reliable for the non-recirculating flows, but all bounds otherwise accurately predict the observed convergence.

Figure 7 shows results for a somewhat larger problem, with spatial mesh of 512×512 elements, to compare convergence estimates computed directly on a small problem size with observed convergence on more practical problem sizes. The length of the time domain is set according to the MGRIT coarsening factor such that there are always four levels in the MGRIT hierarchy (or again 100 time points on the coarse grid in the case of two-level) while preserving the previously stated time step to spatial step relationships. Although less tight than above, convergence estimates on the small problem size provide reasonable estimates of the larger problem in many cases, particularly for smaller coarsening factors. The difference in larger coarsening factors is likely because, e.g. Φ^{16} for a 16×16 mesh globally couples elements, while Φ^{16} for a 512×512 mesh remains a fairly sparse matrix. That is, the mode decomposition of Φ^{16} for $n = 16$ is likely a poor representation for $n = 512$.

Finally, we give insight into how the minimum over x is realized in the TAP. Figures 8 and 9 show the GSVD bounds (i.e. φ_{FCF}) as a function of x, for backward Euler and SDIRK3, respectively, and for each of the three problems and diffusion coefficients. A downside of the GSVD bounds in practice is the cost of computing $||(\Psi - \Phi^k)(I - e^{\mathrm{i}x}\Psi)^{-1}||$ for many values of x. As shown, however, for the diffusion-dominated (nearly symmetric) problems, simply choosing $x = 0$ is sufficient. Interestingly, SDIRK3 bounds have almost no dependence on x for any prob-

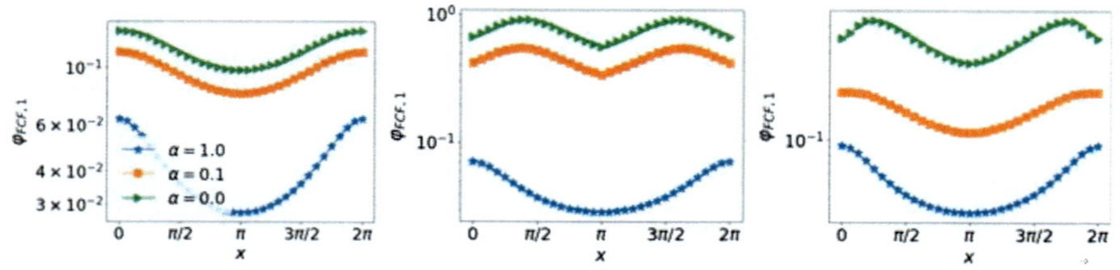

(a) Problem 1 (b) Problem 2 (c) Problem 3

Fig. 8 GSVD bounds ($\varphi_{FCF,1}$) versus choice of x with $n = 16$ and MGRIT coarsening factor 16 using backward Euler for problem 1 (top), 2 (middle), and 3 (bottom) with $\alpha = 1$ (left), 0.01 (middle), and 0.0 (right)

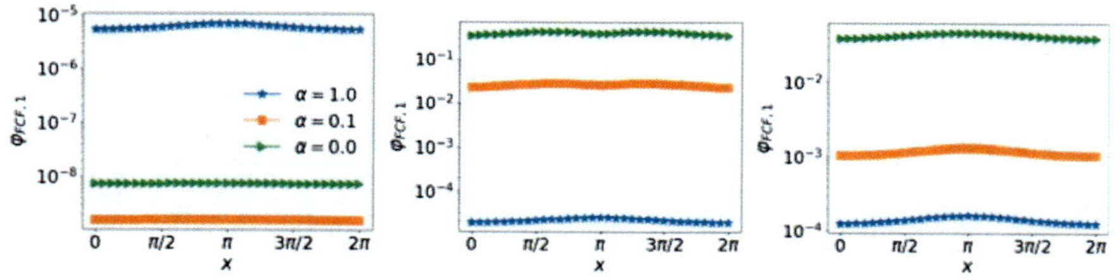

(a) Problem 1 (b) Problem 2 (c) Problem 3

Fig. 9 GSVD bounds ($\varphi_{FCF,1}$) versus choice of x with $n = 16$ and MGRIT coarsening factor 16 using SIDRK3 for problem 1 (top), 2 (middle), and 3 (bottom) with $\alpha = 1$ (left), 0.01 (middle), and 0.0 (right)

lems, while backward Euler bounds tend to have a non-trivial dependence on x (and demonstrate the symmetry in x as discussed in Corollary 1). Nevertheless, accurate bounds for nonsymmetric problems do require sampling a moderate spacing of $x \in [0, \pi]$ to achieve a realistic bound.

4 Conclusion

This paper furthers the theoretical understanding of convergence Parareal and MGRIT. A new, simpler derivation of measuring error and residual propagation operators is provided, which applies to linear time-dependent operators, and which may be a good starting point to develop improved convergence theory for the time-dependent setting. Theory from [17] on spatial operators that are independent of time is then reviewed, and several new results are proven, expanding the understanding of two-level convergence of linear MGRIT and Parareal. Discretizations of the two classical linear hyperbolic PDEs, linear advection (diffusion) and the second-order wave equation, are then analyzed and compared with the theory. Although neither naive implementation yields the rapid convergence desirable in practice, the theory is shown to accurately predict convergence on highly nonsymmetric and hyperbolic operators.

Acknowledgements Los Alamos National Laboratory report number LA-UR-20-28395. This work was supported by Lawrence Livermore National Laboratory under contract B634212, and under the Nicholas C. Metropolis Fellowship from the Laboratory Directed Research and Development program of Los Alamos National Laboratory.

Appendix: Proofs

Proof (*Proof of Theorem* 2) Here, we consider a slightly modified coarsening of points: let the first k points be F-points, followed by a C-point, followed by k F-points, and so on, finishing with a C-point (as opposed to starting with a C-point). This is simply a theoretical tool that permits a cleaner analysis but is not fundamental to the result. Then define the so-called ideal restriction operator, R_{ideal} via

$$\begin{aligned} R_{\text{ideal}} &= \left[-A_{cf}(A_{ff})^{-1} \; I\right] \\ &= \begin{bmatrix} \Phi^{k-1} \; \Phi^{k-2} \ldots \Phi & & & I & \\ & \ddots & & & \ddots & \\ & & \Phi^{k-1} \; \Phi^{k-2} \ldots \Phi & & & I \end{bmatrix}. \end{aligned}$$

Let $\widetilde{P}$ be the orthogonal (column) block permutation matrix such that

$$R_{\text{ideal}}\widetilde{P} = \begin{bmatrix} \Phi^{k-1} \; ... \; \Phi \; I & & \\ & \ddots & \\ & & \Phi^{k-1} \; ... \; \Phi \; I \end{bmatrix} := \begin{bmatrix} \mathbf{W} & & \\ & \ddots & \\ & & \mathbf{W} \end{bmatrix},$$

where $\mathbf{W}$ is the block row matrix $\mathbf{W} = (\Phi^{k-1}, ..., \Phi, I)$. Then, from (4), the norm of residual propagation of MGRIT with F-relaxation is given by

$$\|\mathcal{R}_F\| = \left\|(I - A_\Delta B_\Delta^{-1})R_{\text{ideal}}\right\| = \left\|(I - A_\Delta B_\Delta^{-1})R_{\text{ideal}}\widetilde{P}\right\|,$$

where

$$(I - A_\Delta B_\Delta^{-1})R_{\text{ideal}}\widetilde{P} = \text{diag}\left[\Phi^k - \Psi\right] \begin{bmatrix} \mathbf{0} & & & & \\ I & \mathbf{0} & & & \\ \Psi & I & \mathbf{0} & & \\ \Psi^2 & \Psi & I & \mathbf{0} & \\ \vdots & \ddots & \ddots & \ddots & \ddots \end{bmatrix} \begin{bmatrix} \mathbf{W} & & \\ & \ddots & \\ & & \mathbf{W} \end{bmatrix}$$

$$= \begin{bmatrix} \mathbf{0} & & & & & \\ (\Phi^k - \Psi)\mathbf{W} & \mathbf{0} & & & & \\ (\Phi^k - \Psi)\Psi\mathbf{W} & (\Phi^k - \Psi)\mathbf{W} & \mathbf{0} & & & \\ (\Phi^k - \Psi)\Psi^2\mathbf{W} & (\Phi^k - \Psi)\Psi\mathbf{W} & (\Phi^k - \Psi)\mathbf{W} & \mathbf{0} & & \\ \vdots & \vdots & & \ddots & \ddots & \\ (\Phi^k - \Psi)\Psi^{N_c-2}\mathbf{W} & (\Phi^k - \Psi)\Psi^{N_c-1}\mathbf{W} & \dots & & (\Phi^k - \Psi)\mathbf{W} & \mathbf{0} \end{bmatrix}.$$

Excuse the slight abuse of notation and denote $\mathcal{R}_F := (I - A_\Delta B_\Delta^{-1})R_{\text{ideal}}\widetilde{P}$; that is, ignore the upper zero blocks in $\mathcal{R}_F$. Define a tentative pseudoinverse, $\mathcal{R}_F^\dagger$, as

$$\mathcal{R}_F^\dagger = \begin{bmatrix} \mathbf{0} & \widetilde{\mathbf{W}}^{-1}(\Phi^k - \Psi)^{-1} & & & \\ \mathbf{0} & -\widetilde{\mathbf{W}}^{-1}\Psi(\Phi^k - \Psi)^{-1} & \widetilde{\mathbf{W}}^{-1}(\Phi^k - \Psi)^{-1} & & \\ & & \ddots & \ddots & \\ & & & -\widetilde{\mathbf{W}}^{-1}\Psi(\Phi^k - \Psi)^{-1} & \widetilde{\mathbf{W}}^{-1}(\Phi^k - \Psi)^{-1} \\ & & & \mathbf{0} & \mathbf{0} \end{bmatrix}$$

for some $\widetilde{\mathbf{W}}^{-1}$, and observe that

$$\mathcal{R}_F^\dagger \mathcal{R}_F = \begin{bmatrix} \widetilde{\mathbf{W}}^{-1}\mathbf{W} & & & \\ & \ddots & & \\ & & \widetilde{\mathbf{W}}^{-1}\mathbf{W} & \\ & & & \mathbf{0} \end{bmatrix}.$$

Three of the four properties of a pseudoinverse require that

$$\mathcal{R}_F^\dagger \mathcal{R}_F \mathcal{R}_F^\dagger = \mathcal{R}_F^\dagger, \quad \mathcal{R}_F \mathcal{R}_F^\dagger \mathcal{R}_F = \mathcal{R}_F \quad \text{and} \quad \left(\mathcal{R}_F^\dagger \mathcal{R}_F\right)^* = \mathcal{R}_F^\dagger \mathcal{R}_F.$$

These three properties follow by picking $\widetilde{\mathbf{W}}^{-1}$ such that $\left(\widetilde{\mathbf{W}}^{-1}\mathbf{W}\right)^* = \widetilde{\mathbf{W}}^{-1}\mathbf{W}$, and $\mathbf{W}\widetilde{\mathbf{W}}^{-1}\mathbf{W} = \mathbf{W}$, $\widetilde{\mathbf{W}}^{-1}\mathbf{W}\widetilde{\mathbf{W}}^{-1} = \widetilde{\mathbf{W}}^{-1}$. Notice these are exactly the first three properties of a pseudoinverse of $\mathbf{W}$. To that end, define $\widetilde{\mathbf{W}}^{-1}$ as the pseudoinverse of a full row rank operator,

$$\widetilde{\mathbf{W}}^{-1} = \mathbf{W}^*(\mathbf{W}\mathbf{W}^*)^{-1}.$$

Note that here, $\mathbf{W}\widetilde{\mathbf{W}}^{-1} = I$, and the fourth property of a pseudoinverse for $\mathcal{R}_F$, $\left(\mathcal{R}_F \mathcal{R}_F^\dagger\right)^* = \mathcal{R}_F \mathcal{R}_F^\dagger$, follows immediately.

Recall the maximum singular value of $\mathcal{R}_F$ is given by one over the minimum nonzero singular value of $\mathcal{R}_F^\dagger$, which is equivalent to one over the minimum nonzero singular value of $(\mathcal{R}_F^\dagger)^*\mathcal{R}_F^\dagger$. Following from [17], the minimum nonzero eigenvalue of $(\mathcal{R}_F^\dagger)^*\mathcal{R}_F^\dagger$ is bounded from above by the minimum eigenvalue of a block Toeplitz matrix, with real-valued matrix generating function

$$F(x) = \left(e^{ix}\widetilde{\mathbf{W}}^{-1}\Psi(\Phi^k - \Psi) - \widetilde{\mathbf{W}}^{-1}(\Phi^k - \Psi)\right)\left(e^{ix}\widetilde{\mathbf{W}}^{-1}\Psi(\Phi^k - \Psi) - \widetilde{\mathbf{W}}^{-1}(\Phi^k - \Psi)\right)^*.$$

Let $\lambda_k(A)$ and $\sigma_k(A)$ denote the kth eigenvalue and singular value of some operator, A. Then,

$$\begin{aligned}
\min_{\substack{x\in[0,2\pi],\\ k}} \lambda_k(F(x)) &= \min_{\substack{x\in[0,2\pi],\\ k}} \sigma_k\left(e^{ix}\widetilde{\mathbf{W}}^{-1}\Psi(\Phi^k - \Psi) - \widetilde{\mathbf{W}}^{-1}(\Phi^k - \Psi)\right)^2 \\
&= \min_{\substack{x\in[0,2\pi],\\ \mathbf{v}\neq\mathbf{0}}} \frac{\left\|\left(e^{ix}\widetilde{\mathbf{W}}^{-1}\Psi(\Phi^k - \Psi) - \widetilde{\mathbf{W}}^{-1}(\Phi^k - \Psi)\right)\mathbf{v}\right\|^2}{\|\mathbf{v}\|^2} \\
&= \min_{\substack{x\in[0,2\pi],\\ \mathbf{v}\neq\mathbf{0}}} \frac{\left\|\widetilde{\mathbf{W}}^{-1}(e^{ix}\Psi - I)\mathbf{v}\right\|^2}{\|(\Phi^k - \Psi)\mathbf{v}\|^2} \\
&= \min_{\substack{x\in[0,2\pi],\\ \mathbf{v}\neq\mathbf{0}}} \frac{\left\|(\mathbf{W}\mathbf{W}^*)^{-1/2}(e^{ix}\Psi - I)\mathbf{v}\right\|^2}{\|(\Phi^k - \Psi)\mathbf{v}\|^2}, \\
\sigma_{min}\left(\mathcal{R}_F^\dagger\right)^2 &\leq \min_{\substack{x\in[0,2\pi],\\ \mathbf{v}\neq\mathbf{0}}} \frac{\left\|(\mathbf{W}\mathbf{W}^*)^{-1/2}(e^{ix}\Psi - I)\mathbf{v}\right\|^2}{\|(\Phi^k - \Psi)\mathbf{v}\|} + O(1/N_c).
\end{aligned}$$

Then,[4]

$$\begin{aligned}
\|\mathcal{R}_F\| = \frac{1}{\sigma_{min}\left(\mathcal{R}_F^\dagger\right)} &\geq \frac{1}{\sqrt{\min_{\substack{x\in[0,2\pi],\\ \mathbf{v}\neq\mathbf{0}}} \frac{\left\|(\mathbf{W}\mathbf{W}^*)^{-1/2}(e^{ix}\Psi - I)\mathbf{v}\right\|^2}{\|(\Phi^k - \Psi)\mathbf{v}\|^2} + O(1/N_c)}} \\
&\geq \frac{1}{\min_{\substack{x\in[0,2\pi],\\ \mathbf{v}\neq\mathbf{0}}} \frac{\left\|(\mathbf{W}\mathbf{W}^*)^{-1/2}(e^{ix}\Psi - I)\mathbf{v}\right\|^2}{\|(\Phi^k - \Psi)\mathbf{v}\|^2} + O(1/\sqrt{N_c})} \\
&= \max_{\mathbf{v}\neq\mathbf{0}} \frac{\|(\Phi^k - \Psi)\mathbf{v}\|}{\min_{x\in[0,2\pi]}\left\|(\mathbf{W}\mathbf{W}^*)^{-1/2}(e^{ix}\Psi - I)\mathbf{v}\right\| + O(1/\sqrt{N_c})}.
\end{aligned}$$

The case of $\mathcal{R}_{FCF}$ follows an identical derivation with the modified operator $\widehat{\mathbf{W}} = (\Phi^{2k-1}, \ldots, \Phi^k)$, which follows from the right multiplication by $A_{cf}A_{ff}^{-1}A_{fc}$ in $\mathcal{R}_{FCF} = (I - A_\Delta B_\Delta^{-1})A_{cf}A_{ff}^{-1}A_{fc}R_{\text{ideal}}$, which is effectively a right-diagonal scaling by Φ^k. The cases of error propagation in the ℓ^2-norm follow a similar derivation based on P_{ideal}.

Proof (*Proof of Corollary* 3) The derivations follow a similar path as those in Theorem 2. However, when considering Toeplitz operators defined over the complex scalars (eigenvalues) as opposed to operators, additional results hold. In particular, the previous lower bound (that is, necessary condition) is now a tight bound in norm,

---

[4] More detailed steps for this proof involving the block-Toeplitz matrix and generating function can be found in similar derivations in [13, 15–17].

which follows from a closed form for the eigenvalues of a perturbation to the first or last entry in a tridiagonal Toeplitz matrix [3, 11]. Scalar values also lead to a tighter asymptotic bound, $O(1/N_c)$ as opposed to $O(1/\sqrt{N_c})$, which is derived from the existence of a second-order zero of $F(x) - \min_{y\in[0,2\pi]} F(y)$, when the Toeplitz generating function $F(x)$ is defined over complex scalars as opposed to operators [15]. Analogous derivations for each of these steps can be found in the diagonalizable case in [17], and the steps follow easily when coupled with the pseudoinverse derived in Theorem 2.

Then, noting that

$$\mathcal{W}_F = \sqrt{\sum_{\ell=0}^{k-1}(|\lambda_i|^2)^\ell} = \sqrt{\frac{1-|\lambda_i|^{2k}}{1-|\lambda_i|^2}},$$

$$\mathcal{W}_{FCF} = \sqrt{\sum_{\ell=k}^{2k-1}(|\lambda_i|^2)^\ell} = |\lambda_i|^k\sqrt{\frac{1-|\lambda_i|^{2k}}{(1-|\lambda_i|^2}},$$

and substituting λ_i for Φ and μ_i for Ψ in Theorem 2, the result follows.

References

1. Guillaume Bal. On the Convergence and the Stability of the Parareal Algorithm to Solve Partial Differential Equations. In *Domain Decomposition Methods in Science and Engineering*, pages 425–432. Springer, Berlin, Berlin/Heidelberg, 2005.
2. J. C. Butcher. *Numerical methods for ordinary differential equations*. John Wiley & Sons, Ltd., Chichester, third edition, 2016. With a foreword by J. M. Sanz-Serna.
3. CM Da Fonseca and V Kowalenko. Eigenpairs of a family of tridiagonal matrices: three decades later. *Acta Mathematica Hungarica*, pages 1–14.
4. V A Dobrev, Tz V Kolev, N A Petersson, and J B Schroder. Two-level Convergence Theory for Multigrid Reduction in Time (MGRIT). *SIAM Journal on Scientific Computing*, 39(5):S501–S527, 2017.
5. R D Falgout, S Friedhoff, Tz V Kolev, S P MacLachlan, and J B Schroder. Parallel Time Integration with Multigrid. *SIAM Journal on Scientific Computing*, 36(6):C635–C661, 2014.
6. S Friedhoff and B S Southworth. On "optimal" h-independent convergence of Parareal and multigrid-reduction-in-time using Runge-Kutta time integration. *Numerical Linear Algebra with Applications*, page e2301, 2020.
7. M J Gander and Ernst Hairer. Nonlinear Convergence Analysis for the Parareal Algorithm. In *Domain decomposition methods in science and engineering XVII*, pages 45–56. Springer, Berlin, Berlin, Heidelberg, 2008.
8. M J Gander and S Vandewalle. On the Superlinear and Linear Convergence of the Parareal Algorithm. In *Domain Decomposition Methods in Science and Engineering XVI*, pages 291–298. Springer, Berlin, Berlin, Heidelberg, 2007.
9. A Hessenthaler, B S Southworth, D Nordsletten, O Röhrle, R D Falgout, and J B Schroder. Multilevel convergence analysis of multigrid-reduction-in-time. *SIAM Journal on Scientific Computing*, 42(2):A771–A796, 2020.
10. J-L Lions, Y Maday, and G Turinici. Resolution d'EDP par un Schema en Temps "Parareel". *C. R. Acad. Sci. Paris Ser. I Math.*, 332(661–668), 2001.

11. L Losonczi. Eigenvalues and eigenvectors of some tridiagonal matrices. *Acta Mathematica Hungarica*, 60(3–4):309–322, 1992.
12. T A Manteuffel, K J Ressel, and G Starke. A boundary functional for the least-squares finite-element solution of neutron transport problems. *SIAM Journal on Numerical Analysis*, 37(2):556–586, 1999.
13. M Miranda and P Tilli. Asymptotic Spectra of Hermitian Block Toeplitz Matrices and Preconditioning Results. *SIAM Journal on Matrix Analysis and Applications*, 21(3):867–881, January 2000.
14. Daniel Ruprecht. Convergence of Parareal with Spatial Coarsening. *PAMM*, 14(1):1031–1034, December 2014.
15. S Serra. On the Extreme Eigenvalues of Hermitian (Block) Toeplitz Matrices. *Linear Algebra and its Applications*, 270(1–3):109–129, February 1998.
16. S Serra. Spectral and Computational Analysis of Block Toeplitz Matrices Having Nonnegative Definite Matrix-Valued Generating Functions. *BIT*, 39(1):152–175, March 1999.
17. B S Southworth. Necessary Conditions and Tight Two-level Convergence Bounds for Parareal and Multigrid Reduction in Time. *SIAM J. Matrix Anal. Appl.*, 40(2):564–608, 2019.
18. S-L Wu and T Zhou. Convergence Analysis for Three Parareal Solvers. *SIAM Journal on Scientific Computing*, 37(2):A970–A992, January 2015.
19. Shu-Lin Wu. Convergence analysis of some second-order parareal algorithms. *IMA Journal of Numerical Analysis*, 35(3):1315–1341, 2015.
20. XBraid: Parallel multigrid in time. http://llnl.gov/casc/xbraid.
21. I N Zwaan and M E Hochstenbach. Generalized davidson and multidirectional-type methods for the generalized singular value decomposition. arXiv preprint arXiv:1705.06120, 2017.

A Parallel Algorithm for Solving Linear Parabolic Evolution Equations

Raymond van Venetië and Jan Westerdiep

Abstract We present an algorithm for the solution of a simultaneous space-time discretization of linear parabolic evolution equations with a symmetric differential operator in space. Building on earlier work, we recast this discretization into a Schur complement equation whose solution is a quasi-optimal approximation to the weak solution of the equation at hand. Choosing a tensor-product discretization, we arrive at a remarkably simple linear system. Using wavelets in time and standard finite elements in space, we solve the resulting system in linear complexity on a single processor, and in polylogarithmic complexity when parallelized in both space and time. We complement these theoretical findings with large-scale parallel computations showing the effectiveness of the method.

Keywords Parabolic PDEs · Space-time variational formulations · Optimal preconditioning · Parallel algorithms · Massively parallel computing

1 Introduction

This paper deals with solving parabolic evolution equations in a time-parallel fashion using tensor-product discretizations. Time-parallel algorithms for solving parabolic evolution equations have become a focal point following the enormous increase in parallel computing power. Spatial parallelism is a ubiquitous component in large-

Supplementary material: Source code is available at [30].

R. van Venetië · J. Westerdiep (✉)
Korteweg–de Vries (KdV) Institute for Mathematics, University of Amsterdam, PO Box 94248, 1090 GE Amsterdam, The Netherlands
e-mail: j.h.westerdiep@uva.nl

R. van Venetië
e-mail: r.vanvenetie@uva.nl

© The Author(s), under exclusive license to Springer Nature Switzerland AG 2021

B. Ong et al. (eds.), *Parallel-in-Time Integration Methods*, Springer Proceedings in Mathematics & Statistics 356, https://doi.org/10.1007/978-3-030-75933-9_2

scale computations, but when spatial parallelism is exhausted, parallelization of the time axis is of interest.

Time-stepping methods first discretize the problem in space, and then solve the arising system of coupled ODEs sequentially, immediately revealing a primary source of difficulty for time-parallel computation.

Alternatively, one can solve simultaneously in space *and* time. Originally introduced in [3, 4], these space-time methods are very flexible: some can guarantee quasi-best approximations, meaning that their error is proportional to that of the best approximation from the trial space [1, 7, 10, 24], or drive adaptive routines [20, 23]. Many are especially well suited for time-parallel computation [12, 17]. Since the first significant contribution to time-parallel algorithms [18] in 1964, many methods suitable for parallel computation have surfaced; see the review [11].

Parallel complexity. The (serial) complexity of an algorithm measures asymptotic runtime on a single processor in terms of the input size. *Parallel complexity* measures asymptotic runtime given *sufficiently many* parallel processors having access to a shared memory, i.e. assuming there are no communication costs.

In the current context of tensor-product discretizations of parabolic PDEs, we denote with N_t and $N_{\mathbf{x}}$ the number of unknowns in time and space, respectively.

The Parareal method [15] aims at time-parallelism by alternating a serial coarse-grid solve with fine-grid computations in parallel. This way, each iteration has a time-parallel complexity of $\mathcal{O}(\sqrt{N_t} N_{\mathbf{x}})$, and combined with parallel multigrid in space, a parallel complexity of $\mathcal{O}(\sqrt{N_t} \log N_{\mathbf{x}})$. The popular MGRIT algorithm extends these ideas to multiple levels in time; cf. [9].

Recently, Neumüller and Smears proposed an iterative algorithm that uses a Fast Fourier Transform in time. Each iteration runs serially in $\mathcal{O}(N_t \log(N_t) N_{\mathbf{x}})$ and parallel in time, in $\mathcal{O}(\log(N_t) N_{\mathbf{x}})$. By also incorporating parallel multigrid in space, its parallel runtime may be reduced to $\mathcal{O}(\log N_t + \log N_{\mathbf{x}})$.

Our contribution. In this paper, we study a variational formulation introduced in [27] which was based on work by Andreev [1, 2]. Recently, in [28, 29], we studied this formulation in the context of space-time adaptivity and its efficient implementation in serial and on shared-memory parallel computers. The current paper instead focuses on its massively parallel implementation and time-parallel performance.

Our method has remarkable similarities with the approach of [17], and the most essential difference is the substitution of their Fast Fourier Transform by our Fast Wavelet Transform. The strengths of both methods include a solid inf-sup theory that enables quasi-optimal approximate solutions from the trial space, ease of implementation, and excellent parallel performance in practice.

Our method has another strength: based on a wavelet transform, for fixed algebraic tolerance, it runs serially in linear complexity. Parallel-in-time, it runs in complexity $\mathcal{O}(\log(N_t) N_{\mathbf{x}})$; parallel in *space and time*, in $\mathcal{O}(\log(N_t N_{\mathbf{x}}))$. Moreover, when solving to an algebraic error proportional to the discretization error, incorporating a *nested iteration* (cf. [13, Chap. 5]) results in complexities $\mathcal{O}(N_t N_{\mathbf{x}})$, $\mathcal{O}(\log(N_t) N_{\mathbf{x}})$, and $\mathcal{O}(\log^2(N_t N_{\mathbf{x}}))$, respectively. This is on par with best-known results on parallel complexity for elliptic problems; see also [5].

Organization of this paper. In Sect. 2, we formally introduce the problem, derive a saddle-point formulation, and provide sufficient conditions for quasi-optimality of discrete solutions. In Sect. 3, we detail the efficient computation of these discrete solutions. In Sect. 4 we take a concrete example—the reaction-diffusion equation—and analyze the serial and parallel complexity of our algorithm. In Sect. 5, we test these theoretical findings in practice. We conclude in Sect. 6.

Notations. For normed linear spaces U and V, in this paper for convenience over $\mathbb{R}$, $\mathcal{L}(U, V)$ will denote the space of bounded linear mappings $U \to V$ endowed with the operator norm $\|\cdot\|_{\mathcal{L}(U,V)}$. The subset of invertible operators in $\mathcal{L}(U, V)$ with inverses in $\mathcal{L}(V, U)$ will be denoted as $\mathcal{L}\mathrm{is}(U, V)$.

Given a finite-dimensional subspace U^δ of a normed linear space U, we denote the trivial embedding $U^\delta \to U$ by E_U^δ. For a basis Φ^δ—viewed formally as a column vector—of U^δ, we define the *synthesis operator* as

$$\mathcal{F}_{\Phi^\delta} : \mathbb{R}^{\dim U^\delta} \to U^\delta : \mathbf{c} \mapsto \mathbf{c}^\top \Phi^\delta =: \sum_{\phi \in \Phi^\delta} c_\phi \phi.$$

Equip $\mathbb{R}^{\dim U^\delta}$ with the Euclidean inner product and identify $(\mathbb{R}^{\dim U^\delta})'$ with $\mathbb{R}^{\dim U^\delta}$ using the corresponding Riesz map. We find the adjoint of $\mathcal{F}_{\Phi^\delta}$, the *analysis operator*, to satisfy

$$(\mathcal{F}_{\Phi^\delta})' : (U^\delta)' \to \mathbb{R}^{\dim U^\delta} : f \mapsto f(\Phi^\delta) := [f(\phi)]_{\phi \in \Phi^\delta}.$$

For quantities f and g, by $f \lesssim g$, we mean that $f \leq C \cdot g$ with a constant that does not depend on parameters that f and g may depend on. By $f \eqsim g$, we mean that $f \lesssim g$ and $g \lesssim f$. For matrices $\mathbf{A}$ and $\mathbf{B} \in \mathbb{R}^{N \times N}$, by $\mathbf{A} \eqsim \mathbf{B}$, we will denote *spectral equivalence*, i.e. $\mathbf{x}^\top \mathbf{A}\mathbf{x} \eqsim \mathbf{x}^\top \mathbf{B}\mathbf{x}$ for all $\mathbf{x} \in \mathbb{R}^N$.

2 Quasi-optimal Approximations to the Parabolic Problem

Let V, H be separable Hilbert spaces of functions on some spatial domain such that V is continuously embedded in H, i.e. $V \hookrightarrow H$, with dense compact embedding. Identifying H with its dual yields the Gelfand triple $V \hookrightarrow H \simeq H' \hookrightarrow V'$.

For a.e.

$$t \in I := (0, T),$$

let $a(t; \cdot, \cdot)$ denote a bilinear form on $V \times V$ so that for any $\eta, \zeta \in V$, $t \mapsto a(t; \eta, \zeta)$ is measurable on I, and such that for a.e. $t \in I$,

$$|a(t; \eta, \zeta)| \lesssim \|\eta\|_V \|\zeta\|_V \qquad (\eta, \zeta \in V) \qquad \textit{(boundedness)},$$
$$a(t; \eta, \eta) \gtrsim \|\eta\|_V^2 \qquad (\eta \in V) \qquad \textit{(coercivity)}.$$

With $(A(t)\cdot)(\cdot) := a(t; \cdot, \cdot) \in \mathcal{L}\mathrm{is}(V, V')$, given a forcing function g and initial value u_0, we want to solve the *parabolic initial value problem* of

$$\text{finding } u : I \to V \text{ such that } \quad \begin{cases} \frac{\mathrm{d}u}{\mathrm{d}t}(t) + A(t)u(t) = g(t) \quad (t \in I), \\ u(0) = u_0. \end{cases} \tag{1}$$

2.1 An Equivalent Self-adjoint Saddle-Point System

In a simultaneous space-time variational formulation, the parabolic problem reads as finding u from a suitable space of functions of time and space s.t.

$$(Bw)(v) := \int_I \langle \tfrac{\mathrm{d}w}{\mathrm{d}t}(t), v(t)\rangle_H + a(t; w(t), v(t))\mathrm{d}t = \int_I \langle g(t), v(t)\rangle_H =: g(v)$$

for all v from another suitable space of functions of time and space. One possibility to enforce the initial condition is by testing against additional test functions.

Theorem 1 ([22]) *With $X := L_2(I; V) \cap H^1(I; V')$, $Y := L_2(I; V)$, we have*

$$\begin{bmatrix} B \\ \gamma_0 \end{bmatrix} \in \mathcal{L}\mathrm{is}(X, Y' \times H),$$

where for $t \in \bar{I}$, $\gamma_t : u \mapsto u(t, \cdot)$ denotes the trace map. In other words,

$$\textit{finding } u \in X \textit{ s.t. } \quad (Bu, \gamma_0 u) = (g, u_0) \quad \textit{given} \quad (g, u_0) \in Y' \times H \tag{2}$$

is a well-posed simultaneous space-time variational formulation of (1).

We define $A \in \mathcal{L}\mathrm{is}(Y, Y')$ and $\partial_t \in \mathcal{L}\mathrm{is}(X, Y')$ as

$$(Au)(v) := \int_I a(t; u(t), v(t))\mathrm{d}t, \quad \text{and} \quad \partial_t := B - A.$$

Following [27], we assume that A is *symmetric*. We can reformulate (2) as the self-adjoint saddle point problem

$$\text{finding } (v, \sigma, u) \in Y \times H \times X \text{ s.t. } \begin{bmatrix} A & 0 & B \\ 0 & \mathrm{Id} & \gamma_0 \\ B' & \gamma_0' & 0 \end{bmatrix} \begin{bmatrix} v \\ \sigma \\ u \end{bmatrix} = \begin{bmatrix} g \\ u_0 \\ 0 \end{bmatrix}. \tag{3}$$

By taking a Schur complement w.r.t. the H-block, we can reformulate this as

$$\text{finding } (v, u) \in Y \times X \text{ s.t. } \begin{bmatrix} A & B \\ B' & -\gamma_0'\gamma_0 \end{bmatrix} \begin{bmatrix} v \\ u \end{bmatrix} = \begin{bmatrix} g \\ -\gamma_0' u_0 \end{bmatrix}. \tag{4}$$

We equip Y and X with *"energy"-norms*

$$\|\cdot\|_Y^2 := (A\cdot)(\cdot), \quad \|\cdot\|_X^2 := \|\partial_t\cdot\|_{Y'}^2 + \|\cdot\|_Y^2 + \|\gamma_T\cdot\|_H^2,$$

which are equivalent to the canonical norms on Y and X.

2.2 Uniformly Quasi-optimal Galerkin Discretizations

Our numerical approximations will be based on the saddle-point formulation (4). Let $(Y^\delta, X^\delta)_{\delta\in\Delta}$ be a collection of closed subspaces of $Y \times X$ satisfying

$$X^\delta \subset Y^\delta, \quad \partial_t X^\delta \subset Y^\delta \quad (\delta \in \Delta), \tag{5}$$

and

$$1 \geq \gamma_\Delta := \inf_{\delta\in\Delta} \inf_{0\neq u\in X^\delta} \sup_{0\neq v\in Y^\delta} \frac{(\partial_t u)(v)}{\|\partial_t u\|_{Y'}\|v\|_Y} > 0. \tag{6}$$

Remark 2 In [27, Sect. 4], these conditions were verified for X^δ and Y^δ being tensor-products of (locally refined) finite element spaces in time and space. In [28], we relax these conditions to X_t^δ and Y^δ being *adaptive sparse grids*, allowing adaptive refinement locally in space *and* time simultaneously.

For $\delta \in \Delta$, let $(v^\delta, \overline{u}^\delta) \in Y^\delta \times X^\delta$ solve the Galerkin discretization of (4):

$$\begin{bmatrix} E_Y^{\delta\prime} A E_Y^\delta & E_Y^{\delta\prime} B E_X^\delta \\ E_X^{\delta\prime} B' E_Y^\delta & -E_X^{\delta\prime}\gamma_0'\gamma_0 E_X^\delta \end{bmatrix} \begin{bmatrix} v^\delta \\ \overline{u}^\delta \end{bmatrix} = \begin{bmatrix} E_Y^{\delta\prime} g \\ -E_X^{\delta\prime}\gamma_0' u_0 \end{bmatrix}. \tag{7}$$

The solution $(v^\delta, \overline{u}^\delta)$ of (7) exists uniquely, and exhibits *uniform quasi-optimality* in that $\|u - \overline{u}^\delta\|_X \leq \gamma_\Delta^{-1} \inf_{u_\delta\in X^\delta} \|u - u_\delta\|_X$ for all $\delta \in \Delta$.

Instead of solving a matrix representation of (7) using, e.g. preconditioned MINRES, we will opt for a computationally more attractive method. By taking the Schur complement w.r.t. the Y^δ-block in (7), and replacing $(E_Y^{\delta\prime} A E_Y^\delta)^{-1}$ in the resulting formulation by a *preconditioner* K_Y^δ that can be applied cheaply, we arrive at the *Schur complement formulation* of finding $u^\delta \in X^\delta$ s.t.

$$\underbrace{E_X^{\delta\prime}(B' E_Y^\delta K_Y^\delta E_Y^{\delta\prime} B + \gamma_0'\gamma_0) E_X^\delta}_{=:S^\delta} u^\delta = \underbrace{E_X^{\delta\prime}(B' E_Y^\delta K_Y^\delta E_Y^{\delta\prime} g + \gamma_0' u_0)}_{=:f^\delta}. \tag{8}$$

The resulting operator $S^\delta \in \mathcal{L}\mathrm{is}(X^\delta, X^{\delta\prime})$ is self-adjoint and elliptic. Given a self-adjoint operator $K_Y^\delta \in \mathcal{L}(Y^{\delta\prime}, Y^\delta)$ satisfying, for some $\kappa_\Delta \geq 1$,

$$\frac{\big((K_Y^\delta)^{-1}v\big)(v)}{\big(Av\big)(v)} \in [\kappa_\Delta^{-1}, \kappa_\Delta] \quad (\delta \in \Delta, \quad v \in Y^\delta), \tag{9}$$

the solution u^δ of (8) exists uniquely as well. In fact, the following holds.

Theorem 3 ([27, Remark 3.8]) *Take $(Y^\delta \times X^\delta)_{\delta\in\Delta}$ satisfying (5)–(6), and K_Y^δ satisfying (9). Solutions $u^\delta \in X^\delta$ of (8) are uniformly quasi-optimal, i.e.*

$$\|u - u^\delta\|_X \leq \frac{\kappa_\Delta}{\gamma_\Delta} \inf_{u_\delta \in X^\delta} \|u - u_\delta\|_X \quad (\delta \in \Delta).$$

3 Solving Efficiently on Tensor-Product Discretizations

From now on, we assume that $X^\delta := X_t^\delta \otimes X_{\mathbf{x}}^\delta$ and $Y^\delta := Y_t^\delta \otimes Y_{\mathbf{x}}^\delta$ are *tensor-products*, and for ease of presentation, we assume that the spatial discretizations on X^δ and Y^δ coincide, i.e. $X_{\mathbf{x}}^\delta = Y_{\mathbf{x}}^\delta$, reducing (5) to $X_t^\delta \subset Y_t^\delta$ and $\frac{\mathrm{d}}{\mathrm{d}t} X_t^\delta \subset Y_t^\delta$.

We equip X_t^δ with a basis Φ_t^δ, $X_{\mathbf{x}}^\delta$ with $\Phi_{\mathbf{x}}^\delta$, and Y_t^δ with Ξ^δ.

3.1 Construction of K_Y^δ

Define $\mathbf{O} := \langle \Xi^\delta, \Xi^\delta \rangle_{L_2(I)}$ and $\mathbf{A}_{\mathbf{x}} := \langle \Phi_{\mathbf{x}}^\delta, \Phi_{\mathbf{x}}^\delta \rangle_V$. Given $\mathbf{K}_{\mathbf{x}} \eqsim \mathbf{A}_{\mathbf{x}}^{-1}$ uniformly in $\delta \in \Delta$, define

$$\mathbf{K}_Y := \mathbf{O}^{-1} \otimes \mathbf{K}_{\mathbf{x}}.$$

Then, the preconditioner $K_Y^\delta := \mathcal{F}_{\Xi^\delta \otimes \Phi_{\mathbf{x}}^\delta} \mathbf{K}_Y (\mathcal{F}_{\Xi^\delta \otimes \Phi_{\mathbf{x}}^\delta})' \in \mathcal{L}(Y^{\delta'}, Y^\delta)$ satisfies (9); cf. [28, Sect. 5.6.1].

When Ξ^δ is orthogonal, $\mathbf{O}$ is diagonal and can be inverted exactly. For standard finite element bases $\Phi_{\mathbf{x}}^\delta$, suitable $\mathbf{K}_{\mathbf{x}}$ that can be applied efficiently (at cost linear in the discretization size) are provided by symmetric multigrid methods.

3.2 Preconditioning the Schur Complement Formulation

We will solve a matrix representation of (8) with an iterative solver, thus requiring a preconditioner. Inspired by the constructions of [2, 17], we build an *optimal* self-adjoint coercive preconditioner $K_X^\delta \in \mathcal{L}(X^{\delta'}, X^\delta)$ as a wavelet-in-time block-diagonal matrix with multigrid-in-space blocks.

Let U be a separable Hilbert space of functions over some domain. A given collection $\Psi = \{\psi_\lambda\}_{\lambda \in \vee_\Psi}$ is a *Riesz basis* for U when

$$\overline{\operatorname{span}\Psi} = U, \quad \text{and} \quad \|\mathbf{c}\|_{\ell_2(\vee_\Psi)} \eqsim \|\mathbf{c}^\top \Psi\|_U \quad \text{for all} \quad \mathbf{c} \in \ell_2(\vee_\Psi).$$

Thinking of Ψ being a basis of wavelet type, for indices $\lambda \in \vee_\Psi$, its *level* is denoted $|\lambda| \in \mathbb{N}_0$. We call Ψ *uniformly local* when for all $\lambda \in \vee_\Psi$,

$$\operatorname{diam}(\operatorname{supp}\psi_\lambda) \lesssim 2^{-|\lambda|} \text{ and } \#\{\mu \in \vee_\Psi : |\mu| = |\lambda|, |\operatorname{supp}\psi_\mu \cap \operatorname{supp}\psi_\lambda| > 0\} \lesssim 1.$$

Assume $\Sigma := \{\sigma_\lambda : \lambda \in \vee_\Sigma\}$ is a uniformly local Riesz basis for $L_2(I)$ with $\{2^{-|\lambda|}\sigma_\lambda : \lambda \in \vee_\Sigma\}$ Riesz for $H^1(I)$. Writing $w \in X$ as $\sum_{\lambda \in \vee_\Sigma} \sigma_\lambda \otimes w_\lambda$ for some $w_\lambda \in V$, we define the bounded, symmetric, and coercive bilinear form

$$(D_X \sum_{\lambda \in \vee_\Sigma} \sigma_\lambda \otimes w_\lambda)(\sum_{\mu \in \vee_\Sigma} \sigma_\mu \otimes v_\mu) := \sum_{\lambda \in \vee_\Sigma} \langle w_\lambda, v_\lambda \rangle_V + 4^{|\lambda|} \langle w_\lambda, v_\lambda \rangle_{V'}.$$

The operator $D_X^\delta := E_X^{\delta\,\prime} D_X E_X^\delta$ is in $\mathcal{L}\mathrm{is}(X^\delta, X^{\delta\prime})$. Its norm and that of its inverse are bounded uniformly in $\delta \in \Delta$. When $X^\delta = \operatorname{span} \Sigma^\delta \otimes \Phi_{\mathbf{x}}^\delta$ for some $\Sigma^\delta := \{\sigma_\lambda : \lambda \in \vee_{\Sigma^\delta}\} \subset \Sigma$, the matrix representation of D_X^δ w.r.t. $\Sigma^\delta \otimes \Phi_{\mathbf{x}}^\delta$ is

$$(\mathcal{F}_{\Sigma^\delta \otimes \Phi^\delta})' D_X^\delta \mathcal{F}_{\Sigma^\delta \otimes \Phi^\delta} =: \mathbf{D}_X^\delta = \operatorname{blockdiag}[\mathbf{A}_\mathbf{x} + 4^{|\lambda|} \langle \Phi_\mathbf{x}^\delta, \Phi_\mathbf{x}^\delta \rangle_{V'}]_{\lambda \in \vee_{\Sigma^\delta}}.$$

Theorem 4 ([28, Sect. 5.6.2]) *Define* $\mathbf{M}_\mathbf{x} := \langle \Phi_\mathbf{x}^\delta, \Phi_\mathbf{x}^\delta \rangle_H$. *When we have matrices* $\mathbf{K}_j \eqsim (\mathbf{A}_\mathbf{x} + 2^j \mathbf{M}_\mathbf{x})^{-1}$ *uniformly in* $\delta \in \Delta$ *and* $j \in \mathbb{N}_0$, *it follows that*

$$\mathbf{D}_X^{-1} \eqsim \mathbf{K}_X := \operatorname{blockdiag}[\mathbf{K}_{|\lambda|} \mathbf{A}_\mathbf{x} \mathbf{K}_{|\lambda|}]_{\lambda \in \vee_{\Sigma^\delta}}.$$

This yields an optimal preconditioner $K_X^\delta := \mathcal{F}_{\Sigma^\delta \otimes \Phi^\delta} \mathbf{K}_X (\mathcal{F}_{\Sigma^\delta \otimes \Phi^\delta})' \in \mathcal{L}\mathrm{is}(X^{\delta\prime}, X^\delta)$.

In [19], it was shown that under a "full-regularity" assumption, for quasi-uniform meshes, a multiplicative multigrid method yields $\mathbf{K}_j$ satisfying the conditions of Theorem 4, which can moreover be applied in linear time.

3.3 Wavelets-in-Time

The preconditioner $\mathbf{K}_X$ requires X_t^δ to be equipped with a *wavelet* basis Σ^δ, whereas one typically uses a different (single-scale) basis Φ_t^δ on X_t^δ. To bridge this gap, a basis transformation from Σ^δ to Φ_t^δ is required. We define the wavelet transform as $\mathbf{W}_t := (\mathcal{F}_{\Phi_t^\delta})^{-1} \mathcal{F}_{\Sigma^\delta}$.[1]

Define $V_j := \operatorname{span}\{\sigma_\lambda \in \Sigma : |\lambda| \leq j\}$. Equip each V_j with a (single-scale) basis Φ_j, and assume that $\Phi_t^\delta := \Phi_J$ for some J, so that $X_t^\delta := V_J$. Since $V_{j+1} = V_j \oplus$

[1] In literature, this transform is typically called an *inverse wavelet transform*.

span Σ_j, where $\Sigma_j := \{\sigma_\lambda : |\lambda| = j\}$, there exist matrices $\mathbf{P}_j$ and $\mathbf{Q}_j$ such that $\Phi_j^\top = \Phi_{j+1}^\top \mathbf{P}_j$ and $\Psi_j^\top = \Phi_{j+1}^\top \mathbf{Q}_j$, with $\mathbf{M}_j := [\mathbf{P}_j|\mathbf{Q}_j]$ invertible.

Writing $v \in V_J$ in both forms $v = \mathbf{c}_0^\top \Phi_0 + \sum_{j=0}^{J-1} \mathbf{d}_j^\top \Psi_j$ and $v = \mathbf{c}_J^\top \Phi_J$, the basis transformation $\mathbf{W}_t := \mathbf{W}_J$ mapping wavelet coordinates $(\mathbf{c}_0^\top, \mathbf{d}_0^\top, \ldots, \mathbf{d}_{J-1}^\top)$ to single-scale coordinates $\mathbf{c}_J$ satisfies

$$\mathbf{W}_J = \mathbf{M}_{J-1} \begin{bmatrix} \mathbf{W}_{J-1} & \mathbf{0} \\ \mathbf{0} & \mathbf{Id} \end{bmatrix}, \quad \text{and} \quad \mathbf{W}_0 := \mathbf{Id}. \tag{10}$$

Uniform locality of Σ implies *uniform sparsity* of the $\mathbf{M}_j$, i.e. with $\mathcal{O}(1)$ nonzeros per row and column. Then, assuming a geometrical increase in dim V_j in terms of j, which is true in the concrete setting below, matrix-vector products $\mathbf{x} \mapsto \mathbf{W}_t \mathbf{x}$ can be performed (serially) in linear complexity; cf. [26].

3.4 *Solving the System*

The matrix representation of S^δ and f^δ from (8) w.r.t. a basis $\Phi_t^\delta \otimes \Phi_{\mathbf{x}}^\delta$ of X^δ is

$$\mathbf{S} := (\mathcal{F}_{\Phi_t^\delta \otimes \Phi_{\mathbf{x}}^\delta})' S^\delta \mathcal{F}_{\Phi_t^\delta \otimes \Phi_{\mathbf{x}}^\delta} \quad \text{and} \quad \mathbf{f} := (\mathcal{F}_{\Phi_t^\delta \otimes \Phi_{\mathbf{x}}^\delta})' f^\delta.$$

Envisioning an iterative solver, using Sect. 3.2 we have a preconditioner in terms of the wavelet-in-time basis $\Sigma^\delta \otimes \Phi_{\mathbf{x}}^\delta$, with which their matrix representation is

$$\hat{\mathbf{S}} := (\mathcal{F}_{\Sigma^\delta \otimes \Phi_{\mathbf{x}}^\delta})' S^\delta \mathcal{F}_{\Sigma^\delta \otimes \Phi_{\mathbf{x}}^\delta} \quad \text{and} \quad \hat{\mathbf{f}} := (\mathcal{F}_{\Sigma^\delta \otimes \Phi_{\mathbf{x}}^\delta})' f^\delta. \tag{11}$$

These two forms are related: with the wavelet transform $\mathbf{W} := \mathbf{W}_t \otimes \mathbf{Id}_{\mathbf{x}}$, we have $\hat{\mathbf{S}} = \mathbf{W}^\top \mathbf{S} \mathbf{W}$ and $\hat{\mathbf{f}} = \mathbf{W}^\top \mathbf{f}$, and the matrix representation of (8) becomes

$$\text{finding } \mathbf{w} \quad \text{s.t.} \quad \hat{\mathbf{S}} \mathbf{w} = \hat{\mathbf{f}}. \tag{12}$$

We can then recover the solution in single-scale coordinates as $\mathbf{u} = \mathbf{W}\mathbf{w}$.

We use preconditioned conjugate gradients (PCG), with preconditioner $\mathbf{K}_X$, to solve (12). Given an algebraic error tolerance $\epsilon > 0$ and current guess $\mathbf{w}_k$, we monitor $\mathbf{r}_k^\top \mathbf{K}_X \mathbf{r}_k \leq \epsilon^2$, where $\mathbf{r}_k := \hat{\mathbf{f}} - \hat{\mathbf{S}} \mathbf{w}_k$. This data is available within PCG, and constitutes a stopping criterium: with $u_k^\delta := \mathcal{F}_{\Sigma^\delta \otimes \Phi_{\mathbf{x}}^\delta} \mathbf{w}_k \in X^\delta$, we see

$$\mathbf{r}_k^\top \mathbf{K}_X \mathbf{r}_k = (f^\delta - S^\delta u_k^\delta)(K_X^\delta (f^\delta - S^\delta u_k^\delta)) \eqsim \|u^\delta - u_k^\delta\|_X^2 \tag{13}$$

with $\eqsim$ following from [28, (4.12)], so that the algebraic error satisfies $\|u^\delta - u_k^\delta\|_X \lesssim \epsilon$.

4 A Concrete Setting: The Reaction-Diffusion Equation

On a bounded Lipschitz domain $\Omega \subset \mathbb{R}^d$, take $H := L_2(\Omega)$, $V := H_0^1(\Omega)$, and

$$a(t; \eta, \zeta) := \int_\Omega \mathbf{D}\nabla\eta \cdot \nabla\zeta + c\eta\zeta \mathbf{dx}$$

where $\mathbf{D} = \mathbf{D}^\top \in \mathbb{R}^{d\times d}$ is positive definite, and $c \geq 0$.[2] We note that $A(t)$ is symmetric and coercive. W.l.o.g. we take $I := (0, 1)$, i.e. $T := 1$.

Fix $p_t, p_{\mathbf{x}} \in \mathbb{N}$. With $\{\mathcal{T}_I\}$ the family of quasi-uniform partitions of I into subintervals, and $\{\mathcal{T}_\Omega\}$ that of conforming quasi-uniform triangulations of Ω, we define Δ as the collection of pairs $(\mathcal{T}_I, \mathcal{I}_\Omega)$. We construct our trial and test spaces as

$$X^\delta := X_t^\delta \otimes X_{\mathbf{x}}^\delta, \quad Y^\delta := Y_t^\delta \otimes X_{\mathbf{x}}^\delta,$$

where with $\mathbb{P}_p^{-1}(\mathcal{T})$ denoting the space of piecewise degree-p polynomials on $\mathcal{T}$,

$$X_t^\delta := H^1(I) \cap \mathbb{P}_{p_t}^{-1}(\mathcal{T}_I), \quad X_{\mathbf{x}}^\delta := H_0^1(\Omega) \cap \mathbb{P}_{p_{\mathbf{x}}}^{-1}(\mathcal{T}_\Omega), \quad Y_t^\delta := \mathbb{P}_{p_t}^{-1}(\mathcal{T}_I).$$

These spaces satisfy condition (5), with coinciding spatial discretizations on X^δ and Y^δ. For this choice of Δ, inf-sup condition (6) follows from [27, Theorem 4.3].

For X_t^δ, we choose Φ_t^δ to be the Lagrange basis of degree p_t on $\mathcal{T}_I$; for $X_{\mathbf{x}}^\delta$, we choose $\Phi_{\mathbf{x}}^\delta$ to be that of degree $p_{\mathbf{x}}$ on $\mathcal{T}_\Omega$. An orthogonal basis Ξ^δ for Y_t^δ may be built as piecewise shifted Legendre polynomials of degree p_t w.r.t. $\mathcal{T}_I$.

For $p_t = 1$, one finds a suitable wavelet basis Σ in [25]. For $p_t > 1$, one can either split the system into lowest and higher order parts and perform the transform on the lowest order part only, or construct higher order wavelets directly; cf. [8].

Owing to the tensor-product structure of X^δ and Y^δ and of the operators A and ∂_t, the matrix representation of our formulation becomes remarkably simple.

Lemma 5 *Define* $\mathbf{g} := (\mathcal{F}_{\Xi^\delta \otimes \Phi_{\mathbf{x}}^\delta})' g$, $\mathbf{u}_0 := \Phi_t^\delta(0) \otimes \langle u_0, \Phi_{\mathbf{x}}^\delta \rangle_{L_2(\Omega)}$, *and*

$$\begin{aligned} \mathbf{T} &:= \langle \tfrac{\mathrm{d}}{\mathrm{d}t}\Phi_t^\delta, \Xi^\delta \rangle_{L_2(I)}, & \mathbf{N} &:= \langle \Phi_t^\delta, \Xi^\delta \rangle_{L_2(I)}, \\ \Gamma_0 &:= \Phi_t^\delta(0)[\Phi_t^\delta(0)]^\top, & \mathbf{M}_{\mathbf{x}} &:= \langle \Phi_{\mathbf{x}}^\delta, \Phi_{\mathbf{x}}^\delta \rangle_{L_2(\Omega)}, \\ \mathbf{A}_{\mathbf{x}} &:= \langle \mathbf{D}\nabla\Phi_{\mathbf{x}}^\delta, \nabla\Phi_{\mathbf{x}}^\delta \rangle_{L_2(\Omega)} + c\mathbf{M}_{\mathbf{x}}, & \mathbf{B} &:= \mathbf{T} \otimes \mathbf{M}_{\mathbf{x}} + \mathbf{N} \otimes \mathbf{A}_{\mathbf{x}}. \end{aligned}$$

With $\mathbf{K}_Y := \mathbf{O}^{-1} \otimes \mathbf{K}_{\mathbf{x}}$ *from Sect. 3.1, we can write* $\mathbf{S}$ *and* $\mathbf{f}$ *from Sect. 3.4 as*

$$\mathbf{S} = \mathbf{B}^\top \mathbf{K}_Y \mathbf{B} + \Gamma_0 \otimes \mathbf{M}_{\mathbf{x}}, \quad \mathbf{f} = \mathbf{B}^\top \mathbf{K}_Y \mathbf{g} + \mathbf{u}_0.$$

Note that $\mathbf{N}$ *and* $\mathbf{T}$ *are non-square,* Γ_0 *is very sparse, and* $\mathbf{T}$ *is bidiagonal.*

In fact, assumption (5) allows us to write $\mathbf{S}$ in an even simpler form.

---

[2] This is easily generalized to variable coefficients, but notation becomes more obtuse.

Lemma 6 *The matrix* $\mathbf{S}$ *can be written as*

$$\mathbf{S} = \mathbf{A}_t \otimes (\mathbf{M_x K_x M_x}) + \mathbf{M}_t \otimes (\mathbf{A_x K_x A_x}) + \mathbf{L}^\top \otimes (\mathbf{M_x K_x A_x}) \\ + \mathbf{L} \otimes (\mathbf{A_x K_x M_x}) + \Gamma_0 \otimes \mathbf{M_x}$$

where

$$\mathbf{L} := \langle \tfrac{\mathrm{d}}{\mathrm{d}t}\Phi_t^\delta, \Phi_t^\delta \rangle_{L_2(I)}, \quad \mathbf{M}_t := \langle \Phi_t^\delta, \Phi_t^\delta \rangle_{L_2(I)}, \quad \mathbf{A}_t := \langle \tfrac{\mathrm{d}}{\mathrm{d}t}\Phi_t^\delta, \tfrac{\mathrm{d}}{\mathrm{d}t}\Phi_t^\delta \rangle_{L_2(I)}.$$

This matrix representation does not depend on Y_t^δ *or* Ξ^δ *at all.*

Proof The expansion of $\mathbf{B} := \mathbf{T} \otimes \mathbf{M_x} + \mathbf{N} \otimes \mathbf{A_x}$ in $\mathbf{S}$ yields a sum of five Kronecker products, one of which is

$$(\mathbf{T}^\top \otimes \mathbf{M_x})\mathbf{K}_Y(\mathbf{T} \otimes \mathbf{A_x}) = (\mathbf{T}^\top \mathbf{O}^{-1}\mathbf{N}) \otimes (\mathbf{M_x K_x A_x}).$$

We will show that $\mathbf{T}^\top \mathbf{O}^{-1}\mathbf{N} = \mathbf{L}^\top$; similar arguments hold for the other terms. Thanks to $X_t^\delta \subset Y_t^\delta$, we can define the trivial embedding $F_t^\delta : X_t^\delta \to Y_t^\delta$. Defining

$$T^\delta : X_t^\delta \to Y_t^{\delta'}, \qquad (T^\delta u)(v) := \langle \tfrac{\mathrm{d}}{\mathrm{d}t} u, v \rangle_{L_2(I)},$$
$$M^\delta : Y_t^\delta \to Y_t^{\delta'}, \qquad (M^\delta u)(v) := \langle u, v \rangle_{L_2(I)},$$

we find $\mathbf{O} = (\mathcal{F}_{\Xi^\delta})' M^\delta \mathcal{F}_{\Xi^\delta}$, $\mathbf{N} = (\mathcal{F}_{\Xi^\delta})' M^\delta F_t^\delta \mathcal{F}_{\Phi_t^\delta}$ and $\mathbf{T} = (\mathcal{F}_{\Xi^\delta})' T^\delta \mathcal{F}_{\Phi_t^\delta}$, so

$$\mathbf{T}^\top \mathbf{O}^{-1}\mathbf{N} = (\mathcal{F}_{\Phi_t^\delta})' T^{\delta'} F_t^\delta \mathcal{F}_{\Phi_t^\delta} = \langle \Phi_t, \tfrac{\mathrm{d}}{\mathrm{d}t}\Phi_t \rangle_{L_2(I)} = \mathbf{L}^\top.$$

□

4.1 Parallel Complexity

The *parallel complexity* of our algorithm is the asymptotic runtime of solving (12) for $\mathbf{u} \in \mathbb{R}^{N_t N_\mathbf{x}}$ in terms of $N_t := \dim X_t^\delta$ and $N_\mathbf{x} := \dim X_\mathbf{x}^\delta$, given sufficiently many parallel processors and assuming no communication cost.

We understand the serial (resp. parallel) cost of a matrix $\mathbf{B}$, denoted $C_\mathbf{B}^s$ (resp. $C_\mathbf{B}^p$), as the asymptotic runtime of performing $\mathbf{x} \mapsto \mathbf{Bx} \in \mathbb{R}^N$ in terms of N, on a single (resp. sufficiently many) processors at no communication cost. For *uniformly sparse* matrices, i.e. with $\mathcal{O}(1)$ nonzeros per row and column, the serial cost is $\mathcal{O}(N)$, and the parallel cost is $\mathcal{O}(1)$ by computing each cell of the output concurrently.

From Theorem 4, we see that $\mathbf{K}_X$ is such that $\kappa_2(\mathbf{K}_X\hat{\mathbf{S}}) \lesssim 1$ uniformly in $\delta \in \Delta$. Therefore, for a given algebraic error tolerance ϵ, we require $\mathcal{O}(\log \epsilon^{-1})$ PCG iterations. Assuming that the parallel cost of matrices dominates that of vector addition and inner products, the parallel complexity of a single PCG iteration is dominated by the cost of applying $\mathbf{K}_X$ and $\hat{\mathbf{S}}$. As $\hat{\mathbf{S}} = \mathbf{W}^\top \mathbf{S}\mathbf{W}$, our algorithm runs in complexity

$$\mathcal{O}(\log \epsilon^{-1}[C^\circ_{\mathbf{K}_X} + C^\circ_{\mathbf{W}^\top} + C^\circ_{\mathbf{S}} + C^\circ_{\mathbf{W}}]) \quad (\circ \in \{s, p\}). \tag{14}$$

Theorem 7 *For fixed algebraic error tolerance* $\epsilon > 0$, *our algorithm runs in*

- *serial complexity* $\mathcal{O}(N_t N_\mathbf{x})$;
- *time-parallel complexity* $\mathcal{O}(\log(N_t) N_\mathbf{x})$;
- *space-time-parallel complexity* $O(\log(N_t N_\mathbf{x}))$.

Proof We absorb the constant factor $\log \epsilon^{-1}$ of (14) into $\mathcal{O}$. We analyze the cost of every matrix separately.

The (inverse) wavelet transform. As $\mathbf{W} = \mathbf{W}_t \otimes \mathbf{Id}_\mathbf{x}$, its serial cost equals $\mathcal{O}(C^s_{\mathbf{W}_t} N_\mathbf{x})$. The choice of wavelet allows performing $\mathbf{x} \mapsto \mathbf{W}_t \mathbf{x}$ at linear serial cost (cf. Sect. 3.3) so that $C^s_\mathbf{W} = \mathcal{O}(N_t N_\mathbf{x})$.

Using (10), we write $\mathbf{W}_t$ as the composition of J matrices, each uniformly sparse and hence at parallel cost $\mathcal{O}(1)$. Because the mesh in time is quasi-uniform, we have $J \eqsim \log N_t$. We find that $C^p_{\mathbf{W}_t} = \mathcal{O}(J) = \mathcal{O}(\log N_t)$ so that the time-parallel cost of $\mathbf{W}$ equals $\mathcal{O}(\log(N_t) N_\mathbf{x})$. By exploiting spatial parallelism as well, we find $C^p_\mathbf{W} = \mathcal{O}(\log N_t)$. Analogous arguments hold for $\mathbf{W}_t^\top$ and $\mathbf{W}^\top$.

The preconditioner. Recall that $\mathbf{K}_X := \text{blockdiag}[\mathbf{K}_{|\lambda|} \mathbf{A}_\mathbf{x} \mathbf{K}_{|\lambda|}]_\lambda$. Since the cost of $\mathbf{K}_j$ is independent of j, we see that

$$C^s_{\mathbf{K}_X} = \mathcal{O}\big(N_t \cdot (2C^s_{\mathbf{K}_j} + C^s_{\mathbf{A}_\mathbf{x}})\big) = \mathcal{O}(2N_t C^s_{\mathbf{K}_j} + N_t N_\mathbf{x}).$$

Implementing the $\mathbf{K}_j$ as typical multiplicative multigrid solvers with linear serial cost, we find $C^s_{\mathbf{K}_X} = \mathcal{O}(N_t N_\mathbf{x})$.

Through temporal parallelism, we can apply each block of $\mathbf{K}_X$ concurrently, resulting in a time-parallel cost of $\mathcal{O}(2C^s_{\mathbf{K}_j} + C^s_{\mathbf{A}_\mathbf{x}}) = \mathcal{O}(N_\mathbf{x})$.

By parallelizing in space as well, we reduce the cost of the uniformly sparse $\mathbf{A}_\mathbf{x}$ to $\mathcal{O}(1)$. The parallel cost of multiplicative multigrid on quasi-uniform triangulations is $\mathcal{O}(\log N_\mathbf{x})$; cf. [16]. It follows that $C^p_{\mathbf{K}_X} = \mathcal{O}(\log N_\mathbf{x})$.

The Schur matrix. Using Lemma 5, we write $\mathbf{S} = \mathbf{B}^\top \mathbf{K}_Y \mathbf{B} + \Gamma_0 \otimes \mathbf{M}_\mathbf{x}$, where $\mathbf{B} = \mathbf{T} \otimes \mathbf{M}_\mathbf{x} + \mathbf{N} \otimes \mathbf{A}_\mathbf{x}$, which immediately reveals that

$$C^s_\mathbf{S} = C^s_{\mathbf{B}^\top} + C^s_{\mathbf{K}_Y} + C^s_\mathbf{B} + C^s_{\Gamma_0} \cdot C^s_\mathbf{M} = \mathcal{O}(N_t N_\mathbf{x} + C^s_{\mathbf{K}_Y}), \quad \text{and}$$
$$C^p_\mathbf{S} = \max\left\{C^p_{\mathbf{B}^\top} + C^p_{\mathbf{K}_Y} + C^p_\mathbf{B},\ C^p_{\Gamma_0} \cdot C^p_\mathbf{M}\right\} = \mathcal{O}(C^p_{\mathbf{K}_Y})$$

because every matrix except $\mathbf{K}_Y$ is uniformly sparse. With arguments similar to the previous paragraph, we see that $\mathbf{K}_Y$ (and hence $\mathbf{S}$) has serial cost $\mathcal{O}(N_t N_\mathbf{x})$, time-parallel cost $\mathcal{O}(N_\mathbf{x})$, and space-time-parallel cost $\mathcal{O}(\log N_\mathbf{x})$. □

4.2 *Solving to Higher Accuracy*

Instead of *fixing* the algebraic error tolerance, maybe more realistic is to desire a solution $\tilde{u}^\delta \in X^\delta$ for which the error is proportional to the discretization error, i.e. $\|u - \tilde{u}^\delta\|_X \lesssim \inf_{u_\delta \in X^\delta} \|u - u_\delta\|_X$.

Assuming that this error decays with a (problem-dependent) rate $s > 0$, i.e. $\inf_{u_\delta \in X^\delta} \|u - u_\delta\|_X \lesssim (N_t N_{\mathbf{x}})^{-s}$, then the same holds for the solution u^δ of (8); cf. Theorem 3. When the algebraic error tolerance decays as $\epsilon \lesssim (N_t N_{\mathbf{x}})^{-s}$, a triangle inequality and (13) show that the error of our solution $\tilde{u}^\delta$ obtained by PCG decays at rate s too.

In this case, $\log \epsilon^{-1} = \mathcal{O}(\log(N_t N_{\mathbf{x}}))$. From (14) and the proof of Theorem 7, we find our algorithm to run in superlinear serial complexity $\mathcal{O}(N_t N_{\mathbf{x}} \log(N_t N_{\mathbf{x}}))$, time-parallel complexity $\mathcal{O}(\log^2(N_t)\log(N_{\mathbf{x}})N_{\mathbf{x}})$, and polylogarithmic complexity $\mathcal{O}(\log^2(N_t N_{\mathbf{x}}))$ parallel in space and time.

For elliptic PDEs, algorithms are available that offer quasi-optimal solutions, serially in linear complexity $\mathcal{O}(N_{\mathbf{x}})$—the cost of a serial solve to *fixed* algebraic error—and in parallel in $\mathcal{O}(\log^2 N_{\mathbf{x}})$, by combining a *nested iteration* with parallel multigrid; cf. [13, Chap. 5] and [5].

In [14], the question is posed whether "good serial algorithms for parabolic PDEs are intrinsically as parallel as good serial algorithms for elliptic PDEs", basically asking if the lower bound of $\mathcal{O}(\log^2(N_t N_{\mathbf{x}}))$ can be attained by an algorithm that runs serially in $\mathcal{O}(N_t N_{\mathbf{x}})$; see [32, Sect. 2.2] for a formal discussion.

Nested iteration drives down the serial complexity of our algorithm to a linear $\mathcal{O}(N_t N_{\mathbf{x}})$, and also improves the time-parallel complexity to $\mathcal{O}(\log(N_t)N_{\mathbf{x}})$.[3] This is on par with the best-known results for elliptic problems, so we answer the question posed in [14] in the affirmative.

5 Numerical Experiments

We take the simple heat equation, i.e. $D = \mathbf{Id_x}$ and $c = 0$. We select $p_t = p_{\mathbf{x}} = 1$, i.e. lowest order finite elements in space and time. We will use the three-point wavelet introduced in [25].

We implemented our algorithm in Python using the open source finite element library `NGSolve` [21] for meshing and discretization of the bilinear forms in space and time, `MPI` through `mpi4py` [6] for distributed computations, and `SciPy` [31] for the sparse matrix-vector computations. The source code is available at [30].

[3] Interestingly, nested iteration offers no improvements parallel in space *and* time, with complexity still $\mathcal{O}(\log^2(N_t N_{\mathbf{x}}))$.

Table 1 Computed condition numbers $\kappa_2(\mathbf{K}_X\hat{\mathbf{S}})$. Left: fixed $N_t = 1025$, $N_{\mathbf{x}} = 961$ for varying α. Right: fixed $\alpha = 0.3$ for varying N_t and $N_{\mathbf{x}}$

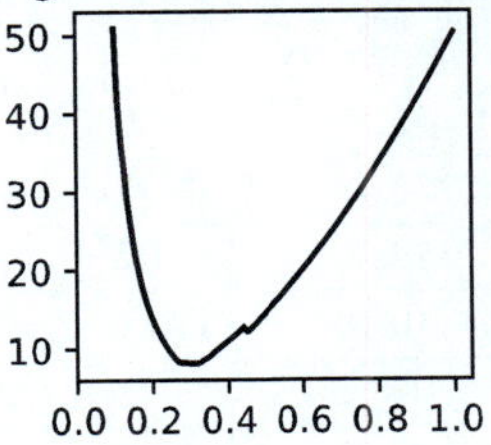

	$N_t = 65$	129	257	513	1 025	2 049	4 097	8 193
$N_{\mathbf{x}} = 49$	6.34	7.05	7.53	7.89	8.15	8.37	8.60	8.78
225	6.33	6.89	7.55	7.91	8.14	8.38	8.57	8.73
961	6.14	6.89	7.55	7.93	**8.15**	8.38	8.57	8.74
3 969	6.14	7.07	7.56	7.87	8.16	8.38	8.57	8.74
16 129	6.14	6.52	7.55	7.86	8.16	8.37	8.57	8.74

5.1 Preconditioner Calibration on a 2D Problem

Our wavelet-in-time, multigrid-in-space preconditioner is optimal: $\kappa_2(\mathbf{K}_X\hat{\mathbf{S}}) \lesssim 1$. Here, we will investigate this condition number quantitatively.

As a model problem, we partition the temporal interval I uniformly into 2^J subintervals. We consider the domain $\Omega := [0, 1]^2$, and triangulate it uniformly into 4^K triangles. We set $N_t := \dim X_t^\delta = 2^J + 1$ and $N_{\mathbf{x}} := \dim X_{\mathbf{x}}^\delta = (2^K - 1)^2$.

We start by using direct inverses $\mathbf{K}_j = (\mathbf{A}_{\mathbf{x}} + 2^j\mathbf{M}_{\mathbf{x}})^{-1}$ and $\mathbf{K}_{\mathbf{x}} = \mathbf{A}_{\mathbf{x}}^{-1}$ to determine the best possible condition numbers. We found that replacing $\mathbf{K}_j$ by $\mathbf{K}_j^\alpha = (\alpha\mathbf{A}_{\mathbf{x}} + 2^j\mathbf{M}_{\mathbf{x}})^{-1}$ for $\alpha = 0.3$ gave better conditioning; see also the left of Table 1. At the right of Table 1, we see that the condition numbers are very robust with respect to spatial refinements, but less so for refinements in time. Still, at $N_t = 16\,129$, we observe a modest $\kappa_2(\mathbf{K}_X\hat{\mathbf{S}})$ of 8.74.

Replacing the direct inverses with multigrid solvers, we found a good balance between speed and conditioning at 2 V-cycles with three Gauss-Seidel smoothing steps per grid. We decided to use these for our experiments.

5.2 Time-Parallel Results

We perform computations on Cartesius, the Dutch supercomputer. Each Cartesius node has 64 GB of memory and 12 cores (at 2 threads per core) running at 2.6 GHz. Using the preconditioner detailed above, we iterate PCG on (12) with $\mathbf{S}$ computed as in Lemma 6, until achieving an algebraic error of $\epsilon = 10^{-6}$; see also Sect. 3.4. For the spatial multigrid solvers, we use 2 V-cycles with three Gauss-Seidel smoothing steps per grid.

Memory-efficient time-parallel implementation. For $\mathbf{X} \in \mathbb{R}^{N_{\mathbf{x}} \times N_t}$, we define $\operatorname{Vec}(\mathbf{X}) \in \mathbb{R}^{N_t N_{\mathbf{x}}}$ as the vector obtained by stacking columns of $\mathbf{X}$ vertically. For memory efficiency, we do not build matrices of the form $\mathbf{B}_t \otimes \mathbf{B}_{\mathbf{x}}$ appearing in Lemma 6 directly, but instead perform matrix-vector products using the identity

Table 2 Strong scaling results for the 2D problem

P	N_t	$N_{\mathbf{x}}$	$N = N_t N_{\mathbf{x}}$	Its	Time (s)	Time/it (s)	CPU-hrs
1–16	16 385	65 025	1 065 434 625	—— Out of memory ——			
32	16 385	65 025	1 065 434 625	16	1224.85	76.55	10.89
64	16 385	65 025	1 065 434 625	16	615.73	38.48	10.95
128	16 385	65 025	1 065 434 625	16	309.81	19.36	11.02
256	16 385	65 025	1 065 434 625	16	163.20	10.20	11.61
512	16 385	65 025	1 065 434 625	16	96.54	6.03	13.73
512	16 385	65 025	1 065 434 625	16	96.50	6.03	13.72
1 024	16 385	65 025	1 065 434 625	16	45.27	2.83	12.88
2 048	16 385	65 025	1 065 434 625	16	20.59	1.29	11.72

$$(\mathbf{B}_t \otimes \mathbf{B}_{\mathbf{x}})\mathrm{Vec}(\mathbf{X}) = \mathrm{Vec}(\mathbf{B}_{\mathbf{x}}(\mathbf{B}_t\mathbf{X}^\top)^\top) = (\mathbf{Id}_t \otimes \mathbf{B}_{\mathbf{x}})\mathrm{Vec}(\mathbf{B}_t\mathbf{X}^\top). \tag{15}$$

Each parallel processor stores only a subset of the temporal degrees-of-freedom, e.g. a subset of columns of $\mathbf{X}$. When $\mathbf{B}_t$ is uniformly sparse, which holds true for all of our temporal matrices, using (15) we can evaluate $(\mathbf{B}_t \otimes \mathbf{B}_{\mathbf{x}})\mathrm{Vec}(\mathbf{X})$ in $\mathcal{O}(C^s_{\mathbf{B}_{\mathbf{x}}})$ operations parallel in time: on each parallel processor, we compute "our" columns of $\mathbf{Y} := \mathbf{B}_t\mathbf{X}^\top$ by receiving the necessary columns of $\mathbf{X}$ from neighbouring processors, and then compute $\mathbf{B}_{\mathbf{x}}\mathbf{Y}^\top$ without communication.

The preconditioner $\mathbf{K}_X$ is block-diagonal, making its time-parallel application trivial. Representing the wavelet transform of Sect. 3.3 as the composition of J Kronecker products allows a time-parallel implementation using the above.

2D problem. We select $\Omega := [0, 1]^2$ with a uniform triangulation $\mathcal{T}_\Omega$, and we triangulate I uniformly into $\mathcal{T}_I$. We select the smooth solution $u(t, x, y) := \exp(-2\pi^2 t) \sin(\pi x) \sin(\pi y)$, so the problem has vanishing forcing data g.

Table 2 details the strong scaling results, i.e. fixing the problem size and increasing the number of processors P. We triangulate I into 2^{14} time slabs, yielding $N_t = 16\,385$ temporal degrees-of-freedom, and Ω into 4^8 triangles, yielding a $X^\delta_{\mathbf{x}}$ of dimension $N_{\mathbf{x}} = 65\,025$. The resulting system contains 1 065 434 625 degrees-of-freedom and our solver reaches the algebraic error tolerance after 16 iterations. In perfect strong scaling, the total number of CPU-hours remains constant. Even at 2 048 processors, we observe a parallel efficiency of around 92.9%, solving this system in a modest 11.7 CPU-hours. Acquiring strong scaling results on a single node was not possible due to memory limitations.

Table 3 details the weak scaling results, i.e. fixing the problem size per processor and increasing the number of processors. In perfect weak scaling, the time per iteration should remain constant. We observe a slight increase in time per iteration on a single node, but when scaling to multiple nodes, we observe a near-perfect parallel efficiency of around 96.7%, solving the final system with 4 278 467 585 degrees-of-freedom in a mere 109 s.

Table 3 Weak scaling results for the 2D problem

	P	N_t	$N_{\mathbf{x}}$	$N = N_t N_{\mathbf{x}}$	Its	Time (s)	Time/it (s)	CPU-hrs
Single node	1	9	261 121	2 350 089	8	33.36	4.17	0.01
	2	17	261 121	4 439 057	11	46.66	4.24	0.03
	4	33	261 121	8 616 993	12	54.60	4.55	0.06
	8	65	261 121	16 972 865	13	65.52	5.04	0.15
	16	129	261 121	33 684 609	13	86.94	6.69	0.39
Multiple nodes	32	257	261 121	67 108 097	14	93.56	6.68	0.83
	64	513	261 121	133 955 073	14	94.45	6.75	1.68
	128	1 025	261 121	267 649 025	14	93.85	6.70	3.34
	256	2 049	261 121	535 036 929	15	101.81	6.79	7.24
	512	4 097	261 121	1 069 812 737	15	101.71	6.78	14.47
	1 024	8 193	261 121	2 139 364 353	16	108.32	6.77	30.81
	2 048	16 385	261 121	4 278 467 585	16	109.59	6.85	62.34

Table 4 Strong scaling results for the 3D problem

P	N_t	$N_{\mathbf{x}}$	$N = N_t N_{\mathbf{x}}$	Its	Time (s)	Time/it (s)	CPU-hrs
1–64	16 385	250 047	4 097 020 095	—— Out of memory ——			
128	16 385	250 047	4 097 020 095	18	3 308.49	174.13	117.64
256	16 385	250 047	4 097 020 095	18	1 655.92	87.15	117.75
512	16 385	250 047	4 097 020 095	18	895.01	47.11	127.29
1 024	16 385	250 047	4 097 020 095	18	451.59	23.77	128.45
2 048	16 385	250 047	4 097 020 095	18	221.12	12.28	125.80

Table 5 Weak scaling results for the 3D problem

P	N_t	$N_{\mathbf{x}}$	$N = N_t N_{\mathbf{x}}$	Its	Time (s)	Time/it (s)	CPU-hrs
16	129	250 047	32 256 063	15	183.65	12.24	0.82
32	257	250 047	64 262 079	16	196.26	12.27	1.74
64	513	250 047	128 274 111	16	197.55	12.35	3.51
128	1 025	250 047	256 298 175	17	210.21	12.37	7.47
256	2 049	250 047	512 346 303	17	209.56	12.33	14.90
512	4 097	250 047	1 024 442 559	17	210.14	12.36	29.89
1 024	8 193	250 047	2 048 635 071	18	221.77	12.32	63.08
2 048	16 385	250 047	4 097 020 095	18	221.12	12.28	125.80

3D problem. We select $\Omega := [0, 1]^3$, and prescribe the solution $u(t, x, y, z) := \exp(-3\pi^2 t)\sin(\pi x)\sin(\pi y)\sin(\pi z)$, so the problem has vanishing forcing data g.

Table 4 shows the strong scaling results. We triangulate I uniformly into 2^{14} time slabs, and Ω uniformly into 8^6 tetrahedra. The arising system has $N = 4\,097\,020\,095$ unknowns, which we solve to tolerance in 18 iterations. The results are comparable to those in two dimensions, albeit a factor two slower at similar problem sizes.

Table 5 shows the weak scaling results for the 3D problem. As in the 2D case, we observe excellent scaling properties and see that the time per iteration is nearly constant.

6 Conclusion

We have presented a framework for solving linear parabolic evolution equations massively in parallel. Based on earlier ideas [2, 17, 27], we found a remarkably simple symmetric Schur complement equation. With a tensor-product discretization of the space-time cylinder using standard finite elements in time and space together with a wavelet-in-time multigrid-in-space preconditioner, we were able to solve the arising systems to fixed accuracy in a uniformly bounded number of PCG steps.

We found that our algorithm runs in linear complexity on a single processor. Moreover, when *sufficiently many* parallel processors are available and communication is free, its runtime scales *logarithmically* in the discretization size. These complexity results translate to a highly efficient algorithm in practice.

The numerical experiments serve as a showcase for the described space-time method and exhibit its excellent time-parallelism by solving a linear system with over 4 billion unknowns in just 109 s, using just over 2000 parallel processors. By incorporating spatial parallelism as well, we expect these results to scale well to much larger problems.

Although performed in the rather restrictive setting of the heat equation discretized using piecewise linear polynomials on uniform triangulations, the parallel framework already allows solving more general linear parabolic PDEs using polynomials of varying degrees on locally refined (tensor-product) meshes. In this more general setting, we envision load balancing to become the main hurdle in achieving good scaling results.

Acknowledgements The authors would like to thank their advisor Rob Stevenson for the many fruitful discussions.

Funding

Both the authors were supported by the Netherlands Organization for Scientific Research (NWO) under contract no. 613.001.652. Computations were performed at the national supercomputer Cartesius under SURF code EINF-459.

References

1. Roman Andreev. Stability of sparse space-time finite element discretizations of linear parabolic evolution equations. *IMA Journal of Numerical Analysis*, 33(1):242–260, 2013.
2. Roman Andreev. Wavelet-In-Time Multigrid-In-Space Preconditioning of Parabolic Evolution Equations. *SIAM Journal on Scientific Computing*, 38(1):A216–A242, 2016.
3. Ivo Babuška and Tadeusz Janik. The h-p version of the finite element method for parabolic equations. Part I. The p-version in time. *Numerical Methods for Partial Differential Equations*, 5(4):363–399, 1989.
4. Ivo Babuška and Tadeusz Janik. The h-p version of the finite element method for parabolic equations. II. The h-p version in time. *Numerical Methods for Partial Differential Equations*, 6(4):343–369, 1990.
5. Achi Brandt. Multigrid solvers on parallel computers. In *Elliptic Problem Solvers*, pages 39–83. Elsevier, 1981.
6. Lisandro Dalcín, Rodrigo Paz, and Mario Storti. MPI for Python. *Journal of Parallel and Distributed Computing*, 65(9):1108–1115, 2005.
7. Denis Devaud and Christoph Schwab. Space–time hp-approximation of parabolic equations. *Calcolo*, 55(3):35, 2018.
8. TJ Dijkema. *Adaptive tensor product wavelet methods for the solution of PDEs*. PhD thesis, Utrecht University, 2009.
9. R. D. Falgout, S. Friedhoff, Tz. V. Kolev, S. P. MacLachlan, and J. B. Schroder. Parallel Time Integration with Multigrid. *SIAM Journal on Scientific Computing*, 36(6):C635–C661, 2014.
10. Thomas Führer and Michael Karkulik. Space-time least-squares finite elements for parabolic equations. 2019. https://doi.org/10.1016/j.camwa.2021.03.004.
11. Martin J. Gander. 50 Years of Time Parallel Time Integration. In *Multiple Shooting and Time Domain Decomposition Methods*, chapter 3, pages 69–113. Springer, Cham, 2015.
12. Martin J. Gander and Martin Neumüller. Analysis of a New Space-Time Parallel Multigrid Algorithm for Parabolic Problems. *SIAM Journal on Scientific Computing*, 38(4):A2173–A2208, 2016.
13. Wolfgang Hackbusch. *Multi-Grid Methods and Applications*, volume 4 of *Springer Series in Computational Mathematics*. Springer Berlin Heidelberg, Berlin, Heidelberg, 1985.
14. G. Horton, S. Vandewalle, and P. Worley. An Algorithm with Polylog Parallel Complexity for Solving Parabolic Partial Differential Equations. *SIAM Journal on Scientific Computing*, 16(3):531–541, 1995.
15. Jacques-Louis Lions, Yvon Maday, and Gabriel Turinici. Résolution d'EDP par un schéma en temps ≪pararéel≫. *Comptes Rendus de l'Académie des Sciences - Series I - Mathematics*, 332(7):661–668, 2001.
16. Oliver A. McBryan, Paul O. Frederickson, Johannes Lindenand, Anton Schüller, Karl Solchenbach, Klaus Stüben, Clemens-August Thole, and Ulrich Trottenberg. Multigrid methods on parallel computers—A survey of recent developments. *IMPACT of Computing in Science and Engineering*, 3(1):1–75, 1991.
17. Martin Neumüller and Iain Smears. Time-parallel iterative solvers for parabolic evolution equations. *SIAM Journal on Scientific Computing*, 41(1):C28–C51, 2019.

18. J. Nievergelt. Parallel methods for integrating ordinary differential equations. *Communications of the ACM*, 7(12):731–733, 1964.
19. Maxim A. Olshanskii and Arnold Reusken. On the Convergence of a Multigrid Method for Linear Reaction-Diffusion Problems. *Computing*, 65(3):193–202, 2000.
20. Nikolaos Rekatsinas and Rob Stevenson. An optimal adaptive tensor product wavelet solver of a space-time FOSLS formulation of parabolic evolution problems. *Advances in Computational Mathematics*, 45(2):1031–1066, 2019.
21. Joachim Schöberl. C++11 Implementation of Finite Elements in NGSolve. Technical report, Institute for Analysis and Scientific Computing, Vienna University of Technology, 2014.
22. Christoph Schwab and Rob Stevenson. Space-time adaptive wavelet methods for parabolic evolution problems. *Mathematics of Computation*, 78(267):1293–1318, 2009.
23. Olaf Steinbach and Huidong Yang. Comparison of algebraic multigrid methods for an adaptive space-time finite-element discretization of the heat equation in 3D and 4D. *Numerical Linear Algebra with Applications*, 25(3):e2143, 2018.
24. Olaf Steinbach and Marco Zank. Coercive space-time finite element methods for initial boundary value problems. *ETNA - Electronic Transactions on Numerical Analysis*, 52:154–194, 2020.
25. Rob Stevenson. Stable three-point wavelet bases on general meshes. *Numerische Mathematik*, 80(1):131–158, 1998.
26. Rob Stevenson. Locally Supported, Piecewise Polynomial Biorthogonal Wavelets on Nonuniform Meshes. *Constructive Approximation*, 19(4):477–508, 2003.
27. Rob Stevenson and Jan Westerdiep. Stability of Galerkin discretizations of a mixed space–time variational formulation of parabolic evolution equations. *IMA Journal of Numerical Analysis*, 2020.
28. Rob Stevenson, Raymond van Venetië, and Jan Westerdiep. A wavelet-in-time, finite element-in-space adaptive method for parabolic evolution equations. 2021.
29. Raymond van Venetië and Jan Westerdiep. Efficient space-time adaptivity for parabolic evolution equations using wavelets in time and finite elements in space. 2021.
30. Raymond van Venetië and Jan Westerdiep. Implementation of: A parallel algorithm for solving linear parabolic evolution equations, 2020. https://doi.org/10.5281/zenodo.4475959.
31. Pauli Virtanen. SciPy 1.0: fundamental algorithms for scientific computing in Python. *Nature Methods*, 17(3):261–272, 2020.
32. Patrick H. Worley. Limits on Parallelism in the Numerical Solution of Linear Partial Differential Equations. *SIAM Journal on Scientific and Statistical Computing*, 12(1):1–35, 1991.

Using Performance Analysis Tools for a Parallel-in-Time Integrator

Does My Time-Parallel Code Do What I Think It Does?

Robert Speck, Michael Knobloch, Sebastian Lührs, and Andreas Gocht

Abstract While many ideas and proofs of concept for parallel-in-time integration methods exists, the number of large-scale, accessible time-parallel codes is rather small. This is often due to the apparent or subtle complexity of the algorithms and the many pitfalls awaiting developers of parallel numerical software. One example of such a time-parallel code is `pySDC`, which implements, among others, the parallel full approximation scheme in space and time (PFASST). Inspired by nonlinear multigrid ideas, PFASST allows to integrate multiple time steps simultaneously using a space-time hierarchy of spectral deferred corrections. In this paper, we demonstrate the application of performance analysis tools to the PFASST implementation `pySDC`. We trace the path we took for this work, show examples of how the tools can be applied, and explain the sometimes surprising findings we encountered. Although focusing only on a single implementation of a particular parallel-in-time integrator, we hope that our results and in particular the way we obtained them are a blueprint for other time-parallel codes.

1 Motivation

With million-way concurrency at hand, the efficient use of modern high-performance computing systems has become one of the key challenges in computational science and engineering. New mathematical concepts and algorithms are needed to fully

R. Speck (✉) · M. Knobloch · S. Lührs
Forschungszentrum Jülich GmbH, Jülich Supercomputing Centre, 52425 Jülich, Germany
e-mail: r.speck@fz-juelich.de

M. Knobloch
e-mail: m.knobloch@fz-juelich.de

S. Lührs
e-mail: s.luehrs@fz-juelich.de

A. Gocht
Center of Information Services and High Performance Computing, Zellescher Weg 12, 01069 Dresden, Germany
e-mail: andreas.gocht@tu-dresden.de

© The Author(s), under exclusive license to Springer Nature Switzerland AG 2021

B. Ong et al. (eds.), *Parallel-in-Time Integration Methods*, Springer Proceedings in Mathematics & Statistics 356, https://doi.org/10.1007/978-3-030-75933-9_3

exploit these massively parallel architectures. For the numerical solution of time-dependent processes, recent developments in the field of parallel-in-time integration have opened new ways to overcome both strong and weak scaling limits of classical, spatial parallelization techniques. In [14], many of these techniques and their properties are presented, while [31] gives an overview of applications of parallel-in-time integration. Furthermore, the community website[1] provides a comprehensive list of references. We refer to these sources for a detailed overview of time-parallel methods and their applications. While many ideas, algorithms, and proofs of concept exist in this domain, the number of actual large-scale time-parallel application codes or even stand-alone parallel-in-time libraries showcasing performance gains is still small. In particular, codes which can deal with parallelization in time as well as in space are rare. At the time of this writing, three main, accessible projects targeting this area are `XBraid`, a C/C++ time-parallel multigrid solver [26], `RIDC`, a C++ implementation of the revisionist integral deferred correction method [32], and at least two different implementations of PFASST, the "parallel full approximation scheme in space and time" [10]. One major PFASST implementation is written in Fortran (`libpfasst`, see [28]), another one in Python (`pySDC`, see [42]).

When running parallel simulations, benchmarks, or just initial tests, one key question is whether the code actually does what it is supposed to do and/or what the developer thinks it does. While this may seem obvious to the developer, complex codes (like PFASST implementations) tend to introduce complex bugs. To avoid these, one may ask for example: How many messages were sent, how many were received? Is there a wait for each non-blocking communication? Are the number of solves/evaluations/iterations reasonable? Moreover, even if the workflow itself is correct and verified, the developer or user may wonder whether the code is as fast as it can be: Is the communication actually non-blocking or blocking, when it should be? Is the waiting time of the processes as expected? Does the algorithm spend reasonable time in certain functions or are there inefficient implementations causing delays? Then, if all runs well, performing comprehensive parameter studies like benchmarking requires a solid workflow management and it can be quite tedious to keep track of what ran where, when, and with what result. In order to address questions like these, advanced performance analysis tools can be used.

The performance analysis tools landscape is manifold. Tools range from node-level analysis tools using hardware counters like LIKWID [44] and PAPI [43] to tools intended for large-scale, complex applications like Scalasca [15]. There are tools developed by the hardware vendors, e.g. Intel VTune [34] or NVIDIA nvprof [5] as well as community-driven open-source tools and tool-sets like Score-P [25], TAU [39], or HPCToolkit [1]. Choosing the right tool depends on the task at hand and of course on the familiarity of the analyst with the available tools.

It is the goal of this paper to present some of these tools and show their capabilities for performance measurements, workflows, and bug detection for time-parallel codes like `pySDC`. Although we will, in the interest of brevity, solely focus on `pySDC` for this paper, our results and in particular the way we obtained them with the different

[1] https://www.parallel-in-time.org.

tools can serve as a blueprint for many other implementations of parallel-in-time algorithms. While there are a lot of studies using these tools for many parallelization strategies, see e.g. [19, 22], and application areas, see e.g. [18, 38], their application in the context of parallel-in-time integration techniques is new. Especially when different parallelization strategies are mixed, these tools can provide invaluable help. We would like to emphasize that this paper is not about the actual results of pySDC, PFASST, or parallel-in-time integration itself (like the application, the parallel speedup, or the time-to-solution), but on the benefits of using performance tools and workflow managers for the development and application of a parallel-in-time integrator. Thus, this paper is meant as a community service to showcase what can be done with a few standard tools from the broad field of HPC performance analysis. One specific challenge in this regard, however, is the programming language of pySDC. Most tools focus on more standard HPC languages like Fortran or C/C++. With the new release of Score-P used for this work, Python codes can now be analyzed as well, as we will show in this paper.

In the next section, we will briefly introduce the PFASST algorithm and describe its implementation in some detail. While the math behind a method may not be relevant for performance tools, understanding the algorithms at least in principle is necessary to give more precise answers to the questions the method developers may have. Section 3 is concerned with a more or less brief and high-level description of the performance analysis tools used for this project. Section 4 describes the endeavor of obtaining reasonable measurements from their application to pySDC, interpreting the results, and learning from them. Section 5 contains a brief summary and an outlook.

2 A Parallel-in-Time Integrator

In this section, we briefly review the collocation problem, being the basis for all problems the algorithm presented here tries to solve in one way or another. Then, spectral deferred corrections (SDC, [9]) are introduced, which lead to the time-parallel integrator PFASST. This section is largely based on [4, 40].

2.1 *Spectral Deferred Corrections*

For ease of notation, we consider a scalar initial value problem on the interval $[t_\ell, t_{\ell+1}]$

$$u_t = f(u), \quad u(t_\ell) = u_0,$$

with $u(t), u_0, f(u) \in \mathbb{R}$. We rewrite this in Picard formulation as

$$u(t) = u_0 + \int_{t_\ell}^{t} f(u(s))ds, \quad t \in [t_\ell, t_{\ell+1}].$$

Introducing M quadrature nodes $\tau_1, \ldots, \tau_M$ with $t_\ell \leq \tau_1 < \ldots < \tau_M = t_{\ell+1}$, we can approximate the integrals from t_ℓ to these nodes τ_m using spectral quadrature like Gauss-Radau or Gauss-Lobatto quadrature such that

$$u_m = u_0 + \Delta t \sum_{j=1}^{M} q_{m,j} f(u_j), \quad m = 1, \ldots, M,$$

where $u_m \approx u(\tau_m)$, $\Delta t = t_{\ell+1} - t_\ell$ and $q_{m,j}$ represent the quadrature weights for the interval $[t_\ell, \tau_m]$ with

$$\Delta t \sum_{j=1}^{M} q_{m,j} f(u_j) \approx \int_{t_\ell}^{\tau_m} f(u(s))ds.$$

We can now combine these M equations into one system of linear or non-linear equations with

$$(\mathbf{I}_M - \Delta t \mathbf{Q} \boldsymbol{f})(\boldsymbol{u}_\ell) = \boldsymbol{u}_0, \tag{1}$$

where $\boldsymbol{u}_\ell = (u_1, \ldots, u_M)^T \approx (u(\tau_1), \ldots, u(\tau_M))^T \in \mathbb{R}^M$, $\boldsymbol{u}_0 = (u_0, \ldots, u_0)^T \in \mathbb{R}^M$, $\mathbf{Q} = (q_{i,j}) \in \mathbb{R}^{M \times M}$ is the matrix gathering the quadrature weights, $\mathbf{I}_M$ is the identity matrix of dimension M, and the vector function $\boldsymbol{f}$ is given by $\boldsymbol{f}(\boldsymbol{u}) = (f(u_1), \ldots, f(u_M))^T \in \mathbb{R}^M$. This system of equations is called the "collocation problem" for the interval $[t_\ell, t_{\ell+1}]$ and it is equivalent to a fully implicit Runge-Kutta method, where the matrix $\mathbf{Q}$ contains the entries of the corresponding Butcher tableau. We note that for $f(u) \in \mathbb{R}^N$, we need to replace $\mathbf{Q}$ by $\mathbf{Q} \otimes \mathbf{I}_N$.

Using SDC, this problem can be solved iteratively and we follow [20, 35, 45] to present SDC as preconditioned Picard iteration for the collocation problem (1). The standard approach to preconditioning is to define an operator which is easy to invert but also close to the operator of the system. One very effective option is the so-called "LU trick", which uses the LU decomposition of $\mathbf{Q}^T$ to define

$$\mathbf{Q}_\Delta = \mathbf{U}^\mathrm{T} \quad \text{for} \quad \mathbf{Q}^T = \mathbf{LU},$$

see [45] for details. With this, we write

$$(\mathbf{I}_M - \Delta t \mathbf{Q}_\Delta \boldsymbol{f})(\boldsymbol{u}_\ell^{k+1}) = \boldsymbol{u}_0 + \Delta t (\mathbf{Q} - \mathbf{Q}_\Delta) \boldsymbol{f}(\boldsymbol{u}_\ell^k) \tag{2}$$

or, equivalently,

$$\boldsymbol{u}_\ell^{k+1} = \boldsymbol{u}_0 + \Delta t \mathbf{Q}_\Delta \boldsymbol{f}(\boldsymbol{u}_\ell^{k+1}) + \Delta t (\mathbf{Q} - \mathbf{Q}_\Delta) \boldsymbol{f}(\boldsymbol{u}_\ell^k) \tag{3}$$

and the operator $\mathbf{I} - \Delta t \mathbf{Q}_\Delta f$ is then called the SDC preconditioner. Writing (3) line by line recovers the classical SDC formulation found in [9].

2.2 Parallel Full Approximation Scheme in Space and Time

We can assemble the collocation problem (1) for multiple time steps, too. Let $\boldsymbol{u}_1, ..., \boldsymbol{u}_L$ be the solution vectors at time steps 1, ..., L and $\vec{\boldsymbol{u}} = (\boldsymbol{u}_1, ..., \boldsymbol{u}_L)^T$ the full solution vector. We define a matrix $\mathbf{H} \in \mathbb{R}^{M \times M}$ such that $\mathbf{H}\boldsymbol{u}_\ell$ provides the initial value for the $\ell + 1$-th time step. Note that this initial value has to be used at all nodes, see the definition of $\boldsymbol{u}_0$ above. The matrix depends on the collocation nodes and if the last node is the right interval boundary, i.e. $\tau_M = t_{\ell+1}$ as it is the case for Gauss-Radau or Gauss-Lobatto nodes, then it is simply given by

$$\mathbf{H} = (0, ..., 0, 1) \otimes (1, ..., 1)^T .$$

Otherwise, $\mathbf{H}$ would contain weights for extrapolation or the collocation formula for the full interval. Note that for $f(u) \in \mathbb{R}^N$, we again need to replace $\mathbf{H}$ by $\mathbf{H} \otimes \mathbf{I}_N$. With this definition, we can assemble the so-called "composite collocation problem" for L time-steps as

$$\mathbf{C}(\vec{\boldsymbol{u}}) := (\mathbf{I}_{LM} - \mathbf{I}_L \otimes \Delta t \mathbf{Q} \boldsymbol{F} - \mathbf{E} \otimes \mathbf{H})(\vec{\boldsymbol{u}}) = \vec{\boldsymbol{u}}_0, \tag{4}$$

with $\vec{\boldsymbol{u}}_0 = (\boldsymbol{u}_0, \mathbf{0}, ..., \mathbf{0})^T \in \mathbb{R}^{LM}$, the vector of vector functions $\vec{\boldsymbol{F}}(\vec{\boldsymbol{u}}) = (\boldsymbol{f}(\boldsymbol{u}_1), ..., \boldsymbol{f}(\boldsymbol{u}_L))^T \in \mathbb{R}^{LM}$ and where the matrix $\mathbf{E} \in \mathbb{R}^{L \times L}$ has ones on the lower sub-diagonal and zeros elsewhere, accounting for the transfer of the solution from one step to another.

For serial time-stepping, each step can be solved after another, i.e. SDC iterations (now called "sweeps") are performed until convergence on $\boldsymbol{u}_1$, move to step 2 via $\mathbf{H}$, do SDC there and so on. In order to introduce parallelism in time, the "parallel full approximation scheme in space in time" (PFASST) makes use of a full approximation scheme (FAS) multigrid approach for solving (4). We present this idea using two levels only, but the algorithm can be easily extended to multiple levels. First, a parallel solver on the fine level and a serial solver on the coarse level are defined as

$$\mathbf{P}_{\text{par}}(\vec{\boldsymbol{u}}) := (\mathbf{I}_{LM} - \mathbf{I}_L \otimes \Delta t \mathbf{Q}_\Delta \boldsymbol{F})(\vec{\boldsymbol{u}}),$$
$$\mathbf{P}_{\text{ser}}(\vec{\boldsymbol{u}}) := (\mathbf{I}_{LM} - \mathbf{I}_L \otimes \Delta t \mathbf{Q}_\Delta \boldsymbol{F} - \mathbf{E} \otimes \mathbf{H})(\vec{\boldsymbol{u}}).$$

Omitting the term $\mathbf{E} \otimes \mathbf{H}$ in $\mathbf{P}_{\text{par}}$ decouples the steps, enabling simultaneous SDC sweeps on each step.

PFASST uses $\mathbf{P}_{\text{par}}$ as smoother on the fine level and $\mathbf{P}_{\text{ser}}$ as an approximate solver on the coarse level. Restriction and prolongation operators $\mathbf{I}_h^H$ and $\mathbf{I}_H^h$ allow transferring information between the fine level (indicated with h) and the coarse level (indicated with H). The approximate solution is then used to correct the solution of the smoother on the finer level. Typically, only two levels are used, although the method is not restricted to this choice. PFASST in its standard implementation allows coarsening in the degrees-of-freedom in space (i.e. use $N/2$ instead of N unknowns per spatial dimension), a reduced collocation rule (i.e. use a different $\mathbf{Q}$ on the coarse level), a less accurate solver in space (for solving (2) on each time step) or even a reduced representation of the problem. The first two strategies directly influence the definition of the restriction and prolongation operators.

Since the right-hand side of the ODE can be a non-linear function, a τ-correction stemming from the FAS is added to the coarse problem. One PFASST iteration then comprises the following steps:

1. Compute τ-correction as

$$\vec{\tau} = \mathbf{C}_H\left(\mathbf{I}_h^H \vec{\boldsymbol{u}}_h^k\right) - \mathbf{I}_h^H \mathbf{C}_h\left(\vec{\boldsymbol{u}}_h^k\right).$$

2. Compute $\vec{\boldsymbol{u}}_H^{k+1}$ from

$$\mathbf{P}_{\text{ser}}(\vec{\boldsymbol{u}}_H^{k+1}) = \vec{\boldsymbol{u}}_{0,H} + \vec{\tau} + (\mathbf{P}_{\text{ser}} - \mathbf{C}_H)\,(\mathbf{I}_h^H \vec{\boldsymbol{u}}_h^k).$$

3. Compute $\vec{\boldsymbol{u}}_h^{k+1/2}$ from

$$\vec{\boldsymbol{u}}_h^{k+1/2} = \vec{\boldsymbol{u}}_h^k + \mathbf{I}_H^h\left(\vec{\boldsymbol{u}}_H^{k+1} - \mathbf{I}_h^H \vec{\boldsymbol{u}}_h^k\right).$$

4. Compute $\vec{\boldsymbol{u}}_h^{k+1}$ from

$$\mathbf{P}_{\text{par}}(\vec{\boldsymbol{u}}_h^{k+1}) = \vec{\boldsymbol{u}}_{0,h} + \left(\mathbf{P}_{\text{par}} - \mathbf{C}_h\right)(\vec{\boldsymbol{u}}_h^{k+1/2}).$$

We note that this "multigrid perspective" [3] does not represent the original idea of PFASST as described in [10, 29]. There, PFASST is presented as a coupling of SDC with the time-parallel method Parareal, augmented by the τ-correction which allows to represent fine-level information on the coarse level.

While conceptually the same, there is a key difference in the implementation of these two representations of PFASST. The workflow of the algorithm is depicted in Fig. 1, showing the original approach in Fig. 1a and the multigrid perspective in Fig. 1b. They differ in the way the fine-level communication is done. As described in [11], under certain conditions, it is possible to introduce overlap of sending/receiving updated values on the fine-level and the coarse-level computations. More precisely, the "window" for finishing fine level communication is as long as two coarse level sweeps: one from the current iteration, one from the predictor which already introduces a lag of later processors (see Fig. 1a). In contrast, the multigrid perspective requires updated fine level values whenever the term $\mathbf{C}_h(\vec{\boldsymbol{u}}_h^k)$ has to be evaluated. This is the case in step 1 and step 2 of the algorithm as presented before. Note that due to the serial nature of step 3, the evaluation of $\mathbf{C}_H(\mathbf{I}_h^H \vec{\boldsymbol{u}}_h^{k+1/2})$ already

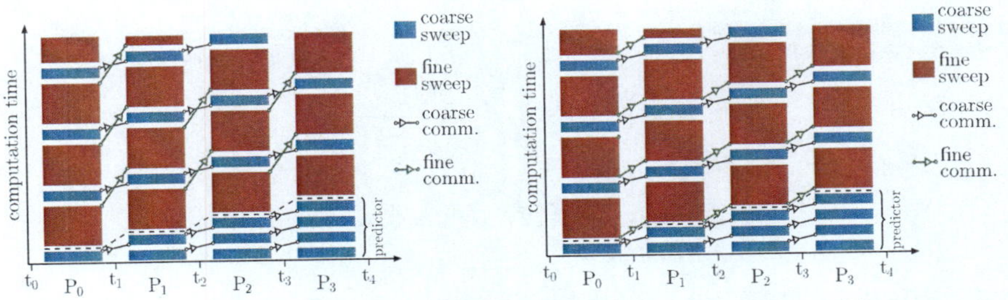

(a) Original algorithm with overlap as described in [10]

(b) Algorithm as described in [3] and implemented in pySDC

Fig. 1 Two slightly different workflows of PFASST, on the left with (theoretically) overlapping fine and coarse communication, on the right with multigrid-like communication

uses the most recent values on the coarse level in both approaches. Therefore, the overlap of communication and computation is somewhat limited: only during the time span of a single coarse-level sweep (introduced by the predictor) the fine-level communication has to finish in order to avoid waiting times (see Fig. 1b). However, the advantage of the multigrid perspective, besides its relative simplicity and ease of notation, is that multiple sweeps on the fine level for speeding up convergence, as shown in [4], are now effectively possible. This is one of the reasons this implementation strategy has been chosen for `pySDC`, while the original Fortran implementation `libpfasst` uses the classical workflow. Yet, while the multigrid perspective may simplify the formal description of the PFASST algorithm, the implementation of PFASST can still be quite challenging.

2.3 pySDC

The purpose of the Python code `pySDC` is to provide a framework for testing, evaluating, and applying different variants of SDC and PFASST without worrying too much about implementation details, communication structures, or lower-level language peculiarities. Users can simply set up an ODE system and run standard versions of SDC or PFASST spending close to no thoughts on the internal structure. In particular, it provides an easy starting point to see whether collocation methods, SDC, and parallel-in-time integration with PFASST are useful for the problem under consideration. Developers, on the other hand, can build on the existing infrastructure to implement new iterative methods or to improve existing methods by overriding any component of `pySDC`, from the main controller and the SDC sweeps to the transfer routines or the way the hierarchy is created. `pySDC`'s main features are [40]:

- available implementations of many variants of SDC, MLSDC, and PFASST,
- many ordinary and partial differential equations already pre-implemented,
- tutorials to lower the bar for new users and developers,

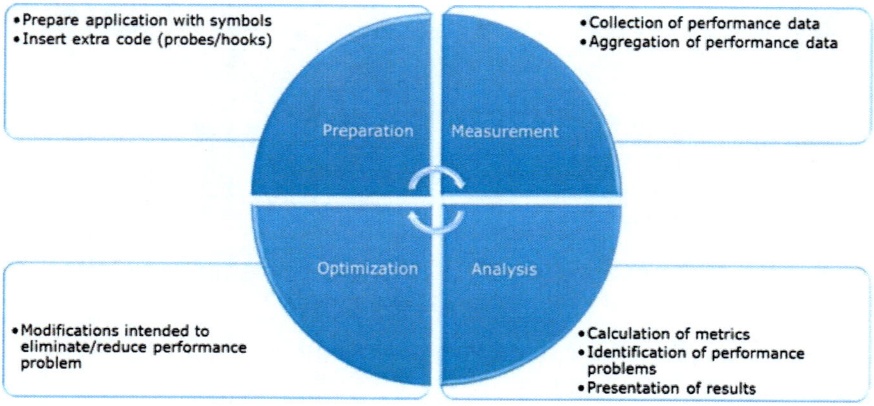

Fig. 2 Performance engineering workflow

- coupling to FEniCS and PETSc, including spatial parallelism for the latter
- automatic testing of new releases, including results of previous publications
- full compatibility with Python 3.6+, runs on desktops and HPC machines

The main website for `pySDC`[2] provides all relevant information, including links to the code repository on Github, the documentation as well as test coverage reports. `pySDC` is also described in much more detail in [40].

The algorithms within `pySDC` are implemented using two "controller" classes. One emulates parallelism in time, while the other one uses `mpi4py` [7] for actual parallelization in the time dimension with the Message Passing Interface (MPI). Both can run the same algorithms and yield the same results, but while the first one is primarily used for theoretical purposes and debugging, the latter makes actual performance tests and time-parallel applications possible.

We will use the MPI-based controller for this paper in order to address the questions posed at the beginning. To do that, a number of HPC tools are available which helps users and developers of HPC software to evaluate the performance of their codes and to speed up their workflows.

3 Performance Analysis Tools

Performance analysis plays a crucial part in the development process of an HPC application. It usually starts with simply timing the computational kernels to see where the time is spent. To access more information and to determine tuning potential, more sophisticated tools are required. The typical performance engineering workflow when using performance analysis tools is an iterative process as depicted in Fig. 2.

[2] https://www.parallel-in-time.org/pySDC.

First, the application needs to be prepared and some hooks to the measurement system need to be added. These can be debug symbols, compiler instrumentation, or even code changes by the user. Then, during the execution of the application, performance data is collected and, if necessary, aggregated. The analysis tools then calculate performance metrics to pinpoint performance problems for the developer. Finally, the hardest part: the developer has to modify the application to eliminate or at least reduce the performance problems found by the tools, ideally without introducing new ones. Unfortunately, tools can only provide little help in this step.

Several performance analysis tools exist, for all kinds of measurement at all possible scales, from a desktop computer to the largest supercomputers in the world. We distinguish two major measurement techniques with different levels of accuracy and overhead:"profiling", which aggregates the performance metrics at runtime and presents statistical results, e.g. how often a function was called and how much time was spend there, and "event-based tracing", where each event of interest, like function enter/exit, messages sent/received, etc. are stored together with a timestamp. Tracing conserves temporal and spatial relationships of events and is the more general measurement technique, as a profile can always be generated from a trace. The main disadvantage of tracing is that trace files can quickly become extremely large (in the order of terabytes) when collecting every event. So usually the first step is a profile to determine the hot-spot of the application, which then is analyzed in detail using tracing to keep trace-size and overhead manageable.

However, performance analysis tools cannot only be used to identify optimization potential but also to assess the execution of the application on a given system with a specific toolchain (compiler, MPI library, etc.), i.e. to answer the question "Is my application doing what I think it is doing?". More often than not the answer to that question is "No", as it was in the case we present in this work. Tools can pinpoint the issues and help to identify possible solutions.

For our analysis, we used the tools of the Score-P ecosystem, which are presented in this section. A similar analysis is possible with other tools as well, e.g. with TAU [39], Paraver [33], or Intels VTune [34].

3.1 Score-P and the Score-P Ecosystem

The Score-P measurement infrastructure [25] is an open-source, highly scalable, and easy-to-use tool suite for profiling, event tracing, and online analysis of HPC applications. It is a community project to replace the measurement systems of several performance analysis tools and to provide common data formats to improve interoperability between different analysis tools built on top of Score-P. Figure 3 shows a schematic overview of the Score-P ecosystem. Most common HPC programming paradigms are supported by Score-P: MPI (via the PMPI interface), OpenMP (via OPARI2, or the OpenMP tools interface (OMPT) [13]) as well as GPU programming with CUDA, OpenACC or OpenCL. Score-P offers three ways to measure application events:

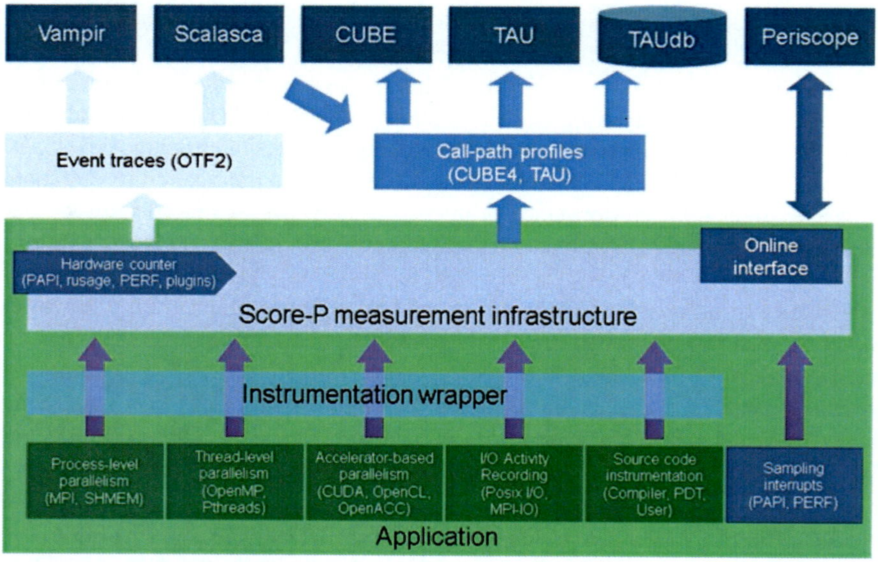

Fig. 3 Overview of the Score-P ecosystem. The green box represents the measurement infrastructure with the various ways of data acquisition. This data is processed by the Score-P measurement infrastructure and stored either aggregated in the CUBE4 profile format or as an event trace in the OTF2 format. On top are the various analysis tools working with these common data formats

1. compiler instrumentation, where compiler interfaces are used to insert calls to the measurement system at each function enter and exit,
2. a user instrumentation API, that enables the application developer to mark specific regions, e.g. kernels, functions or even loops, and
3. a sampling interface which records the state of the program at specific intervals.

All this data is handled in the Score-P measurement core where it can be enriched with hardware counter information from PAPI [43], perf or rusage. Further, Score-P provides a counter plugin interface that enables the user to define its own metric sources. The Score-P measurement infrastructure supports two modes of operation, it can generate event traces in the OTF2 format [12] and aggregated profiles in the CUBE4 format [36].

Usage of Score-P is quite straightforward—the compile and link command have to be prepended by `scorep`, e.g. `mpicc app.c` becomes `scorep mpicc app.c`. However, Score-P can be extensively configured via environment variables, so that Score-P can be used in all analysis steps from a simple call-path profile to a sophisticated tracing experiment enriched with hardware counter information. Listing 3 in Sect. 4.2 will show an example job script where several Score-P options are used.

Score-P Python bindings Traditionally the main programming languages for HPC application development have been C, C++, and Fortran. However, with the advent of high-performance Python libraries in the wake of the rise of AI and deep learning, pure Python HPC applications are now a feasible possibility, as `pySDC` shows. Python has two built-in performance analysis tools, called `profile` and `cProfile`. Though they allow profiling Python code, they do not support as detailed application analyses as Score-P does. Therefore, the Score-P Python bindings have been introduced [17], which allow to profile and trace Python applications using Score-P. This technique can analyze all different kinds of applications that use python, including machine learning workflows. This particular aspect will be described in more detail in another paper.

The bindings use the Python built-in infrastructure that generates events for each enter and exit of a function. It is the same infrastructure that is used by the `profile` tool. As the bindings utilize Score-P itself, the different paradigms listed above can be combined and analyzed even if they are used from within a Python application.

Especially the MPI support of Score-P is of interest, as `pySDC` uses `mpi4py` for parallelization in time. Note that `mpi4py` uses matched probes and receives (`MPI_Mprobe` and `MPI_Mrecv`), which ensures thread safety. However, Score-P did not have support for `Mprobe/Mrecv` in the released version, so we had to switch to a development version of Score-P, where the support was added for this project. Full support for matched communication is expected in an upcoming release of Score-P.

Moreover, as not each function might be of interest for the analysis of an application, it is possible to manually enable and disable the instrumentation or to instrument regions manually, see Listing 4 in Sect. 4.2 for an example. These techniques can be used to control the detail of recorded information and therefore to control the measurement overhead.

3.2 *Cube*

Cube is the performance report explorer for Score-P as well as for Scalasca (see below). The CUBE data model is a 3D performance space consisting of the dimensions (i) performance metric, (ii) call-path, and (iii) system location. Each dimension is represented in the GUI as a tree and shown in three coupled tree browsers, i.e. upon selection of one tree item the other trees are updated. Non-leaf nodes of each tree can be collapsed or expanded to achieve the desired level of granularity. We refer to Fig. 12 for the graphical user interface of Cube. The metrics that are recorded by default contain the time per call, the number of calls to each function, and the bytes transferred in MPI calls. Additional metrics depend on the measurement configuration. The CubeGUI is highly customizable and extendable. It provides a plugin interface to add new analysis capabilities [23] and an integrated domain-specific language called CubePL to manipulate CUBE metrics [37], enabling completely new kinds of analysis.

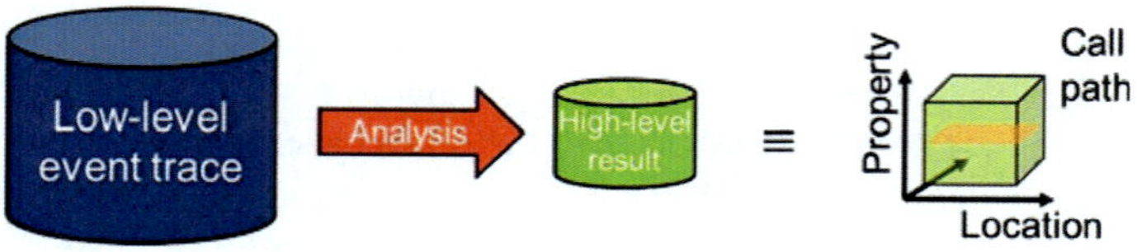

Fig. 4 The Scalasca approach for a scalable parallel trace analysis. The entire trace date is analyzed and only a high-level result is stored in the form of a Cube report

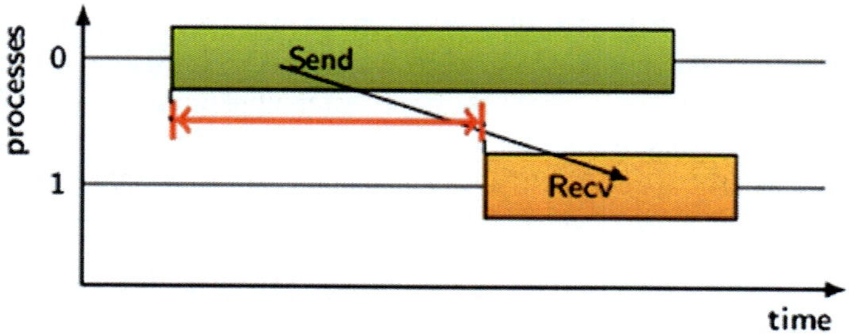

Fig. 5 Example of the Late Receiver pattern as detected by Scalasca. Process 0 post the Send before process 1 posts the Recv. The red arrow indicates waiting time and thus a performance inefficiency

3.3 Scalasca

Scalasca [15] is an automatic analyzer of OTF2 traces generated by Score-P. The idea of Scalasca, as outlined in Fig. 4, is to perform an automatic search for patterns indicating inefficient behavior. The whole low-level trace data is considered and only a high-level result in the form of a CUBE report is generated. This report has the same structure as a Score-P profile report but contains additional metrics for the patterns that Scalasca detected. Scalasca performs three major tasks: (i) an identification of wait states, like the Late Receiver pattern shown in Fig. 5 and their respective root-causes [47], (ii) a classification of the behavior and a quantification of its significance, and (iii) a scalable identification of the critical path of the execution [2]. As Scalasca is primarily targeted at large-scale applications, the analyzer is a parallel program itself, typically running on the same resources as the original application. This enables a unique scalability to the largest machines available [16]. Scalasca offers convenience commands to start the analysis right after measurement in the same job. Unfortunately, this does not work with Python yet, in this case, the analyzer has to be started separately, see line 21 in Listing 3.

3.4 Vampir

Complementary to the automatic trace analysis with Scalasca—and often more intuitive to the user—is a manual analysis with Vampir. Vampir [24] is a powerful trace viewer for OTF2 trace files. In contrast to traditional profile viewers, which only visualize the call hierarchy and function runtimes, Vampir allows the investigation of the whole application flow. Any metrics collected by Score-P, from PAPI or counter plugins, can be analyzed across processes and time with either a timeline or as a heatmap in the Performance Radar. Recently added was the functionality to visualize I/O-events like reads and writes from and to the hard drive [30]. It is possible to zoom into any level of detail, which automatically updated all views and shows the information from the selected part of the trace. Besides opening an OTF2 file directly, Vampir can connect to VampirServer, which uses multiple MPI processes on the remote system to load the traces. This approach improves scalability and removes the necessity to copy the trace file. VampirServer allows the visualization of traces from large-scale application runs with multiple thousand processes. The size of such traces is typically in the order of several Gigabyte.

3.5 JUBE

Managing complex workflows of HPC applications can be a complex and error-prone task and often results in significant amounts of manual work. Application parameters may change at several steps in these workflows. In addition, reproducibility of program results is very important but hard to handle when parametrizations change multiple times during the development process. Usually, application-specific, hardly documented script-based solutions are used to accomplish these tasks.

In contrast, the JUBE benchmarking environment provides a lightweight, command line-based, configurable framework to specify, run, and monitor the parameter handling and the general workflow execution. This allows a faster integration process and easier adoption of necessary workflow mechanics [27].

Parameters are the central JUBE elements and can be used to configure the application, to replace parts of the source code, or to be even used within other parameters. Also, the workflow execution itself is managed through the parameter setup by automatically looping through all available parameter combinations in combination with a dependency-driven step structure. For reproducibility, JUBE also takes care of the directory management to provide a sandbox space for each execution. Finally, JUBE allows to extract relevant patterns from the application output to create a single result overview to combine the input parametrization and the extracted output results.

To port an application workflow into the JUBE framework, its basic compilation (if requested) and execution command steps have to be listed within a JUBE configuration file. To allow the sandbox directory handling, all necessary external files (source codes, input data, and configuration files) have to be listed as well. On top,

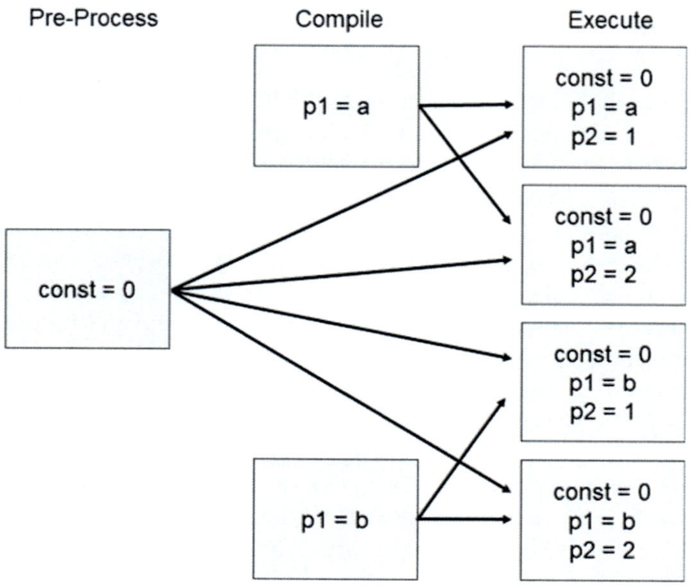

Fig. 6 JUBE workflow example

the user can add the specific parametrization by introducing named key/value pairs. These pairs can either provide a fixed one-to-one key/value mapping or, in case of a parameter study, multiple values can be mapped to the same key. In such a case, JUBE starts to spawn a decision tree, by using every available value combination for a separate program step execution. Figure 6 shows a simple graph example where three different program steps (pre-processing, compile, and execution) are executed in a specific order and three different parameters (`const`, `p1` and `p2`) are defined. Once the parameters are defined, they can be used to substitute parts of the original source files or to directly define certain options within the individual program configuration list. Typically, an application-specific template file is designed to be filled by JUBE parameters afterward. Once the templates and the JUBE configuration file are in place, the JUBE command line tools are used to start the overall workflow execution. JUBE automatically spawns the necessary parameter tree, creates the sandbox directories, and executes the given commands multiple times based on the parameter configuration.

To take care of the typical HPC environment, JUBE also helps with the job submission part by providing a set of job scheduler-specific script templates. This is especially helpful for scaling tests by easily varying the amount of computing devices using a single parameter within the JUBE configuration file. JUBE itself is not aware of the different types of HPC schedulers, therefore, it uses a simple marker file mechanic to recognize if a specific job was finally executed. In Sect. 4.1, we show detailed examples for a configuration file and a jobscript template.

The generic approach of JUBE allows it to easily replace any manual workflow. For example, to use JUBE for an automated performance analysis, using the highlighted performance tools, the necessary Score-P and Scalasca command line options can be directly stored within a parameter, which can then be used during compilation and job submission. After the job execution, even the performance metric extraction can be automated, by converting the profiling data files within an additional performance tool-specific post-processing step into a JUBE parsable output format. This approach allows to easily rerun a specific analysis or even combine performance analysis together with a scaling run, to determine individual metric degradation towards scaling capabilities.

4 Results and Lessons Learned

In the following, we consider the 2D Allen-Cahn equation

$$u_t = \Delta u - \frac{2}{\epsilon^2} u(1-u)(1-2u) \tag{5}$$

$$u(x,0) = \sum_{i=1}^{L} \sum_{j=1}^{L} u_{i,j}(x)$$

with periodic boundary conditions, scaling parameter $\epsilon > 0$ and $x \in \mathbb{R}^N$, $N \in \mathbb{N}$. Note that as a slight abuse of notation $u(x,0)$ is the initial condition for the initial value problem, whereas in Sect. 2.1 u_0 represents the initial value for the individual time slabs. The domain in space $[-L/2, L/2]^2$, $L \in \mathbb{N}$, consists of L^2 patches of size 1×1 and, in each patch, we start with a circle

$$u_{i,j}(x) = \frac{1}{2}\left(1 + \tanh\left(\frac{R_{i,j} - \|x\|}{\sqrt{2}\epsilon}\right)\right)$$

of initial radius $R_{i,j} > 0$ which is chosen randomly between 0.5ϵ and 3ϵ for each patch. For $L = 1$ and this set of parameters, this is precisely the well-known shrinking circle problem, where the dynamics is known and which can be used to verify the simulation [46]. By increasing the parameter L, the simulation domain can be increased without changing the evolution of the simulation fundamentally. For the test shown here, we use $L = 4$, finite differences in space with $N = 576$ and $\epsilon = 0.04$ so that initially about 6 points resolve the interfaces, which have a width of about 7ϵ. We furthermore use $M = 3$ Gauss-Radau nodes and $\Delta t = 0.001 < \epsilon^2$ for the collocation problem and stop the simulation after 24 time steps at $T = 0.024$. We split the right-hand side of (5) and treat the linear diffusion part implicitly using the LU trick [45] and the nonlinear reaction part explicitly using the explicit Euler preconditioner. This has been shown to be the fastest SDC variant in [40] and allows us to use the `mpi4py-fft` library [8] for solving the implicit system, for applying

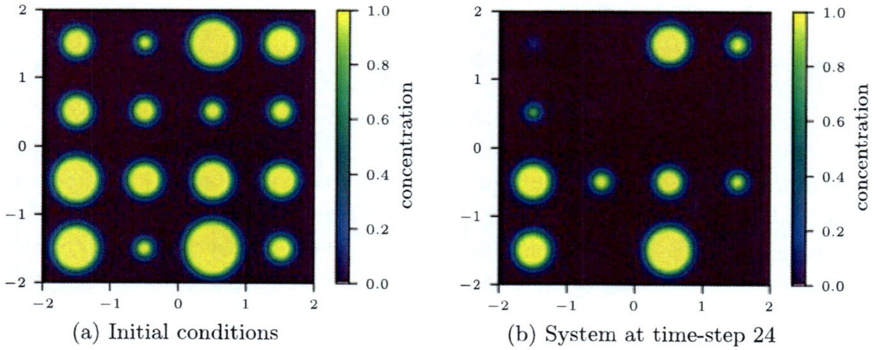

Fig. 7 Evolution of the Allen-Cahn problem used for this analysis

Fig. 8 Time versus number of cores in space and time

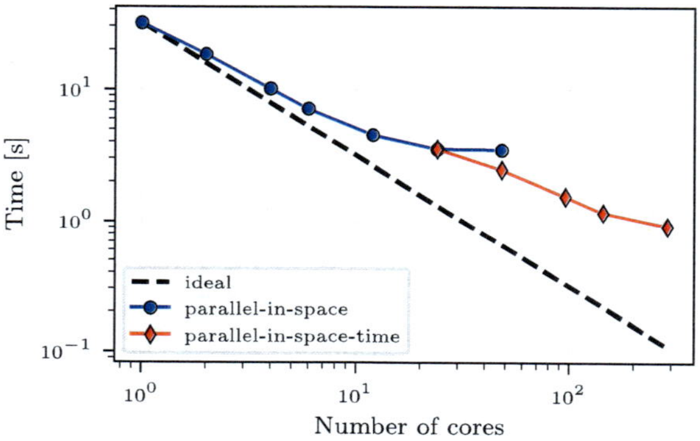

the Laplacian, and for transferring data between coarse and fine levels in space. The iterations are stopped when a residual tolerance of 10^{-8} is reached. For coarsening, only 96 points in space were used on the coarse level and, following [4], 3 sweeps are done on the fine level and 1 on the coarse one. All tests were run on the JURECA cluster at JSC [21] using Python 3.6.8 with the Intel compiler and (unless otherwise stated) Intel MPI. The code can be found in the `projects/Performance` folder of `pySDC` [41]. Figure 7 shows the evolution of the system with $L = 4$ from the initial condition in Fig. 7a to the 24th time step in Fig. 7b.

4.1 Scalability Test with JUBE

In Fig. 8, the scalability of the code in space and time is shown. While spatial parallelization stagnates at about 24 cores, adding temporal parallelism with PFASST allows to use 12 times more processors for an additional speedup of about 4. Note that using even more cores in time increases the runtime again due to a much higher number of iterations. Also, using more than 48 cores in space is not possible due to the size of the problem. We do not consider larger-scale problems and parallelization here, since a detailed performance analysis in this case is currently a work in progress together with the EU Centre of Excellence "Performance Optimisation and Productivity" (POP CoE, see [6] for details).

```xml
<?xml version="1.0" encoding="UTF-8"?>
<jube>
  <benchmark name="pySDC AC scaling test" outpath="bench_run_SPxTP">
    <comment>Scaling test with pySDC</comment>

    <!-- Parameters -->
    <parameterset name="param_set">
      <parameter name="i">0, 1, 2, 3, 4, 5, 6, 7, 8, 9</parameter>
      <parameter name="nnodes" mode="python" type="int">
            [1, 1, 1, 1,  1,  1,  2,  4,  6, 12][$i]</parameter>
      <parameter name="ntasks" mode="python" type="int">
            [1, 2, 4, 6, 12, 24, 24, 24, 24, 24][$i]</parameter>
      <parameter name="space_size" mode="python" type="int">
            $ntasks</parameter>
      <parameter name="mpi" type="str">intel, parastation</parameter>
    </parameterset>

    <!-- Substitute -->
    <substituteset name="substitute">
      <!-- Substitute files -->
      <iofile in="run_pySDC_AC.tmpl" out="run_pySDC_AC.exe" />
      <!-- Substitute commands -->
      <sub source="#NNODES#" dest="$nnodes" />
      <sub source="#NTASKS#" dest="$ntasks" />
      <sub source="#SPACE_SIZE#" dest="$space_size" />
      <sub source="#MPI#" dest="$mpi" />
    </substituteset>

    <!-- Files -->
    <fileset name="files">
      <copy>run_pySDC_AC.tmpl</copy>
      <copy>run_benchmark.py</copy>
    </fileset>

    <!-- Operation -->
    <step name="sub_step">
      <use>param_set</use>    <!-- use existing parameterset -->
      <use>files</use>        <!-- use existing fileset -->
      <use>substitute</use>   <!-- use existing substituteset -->
      <!-- shell command -->
      <do done_file="ready">sbatch run_pySDC_AC.exe</do>
    </step>

    ...
```

Listing 1: XML input file for JUBE running space-parallel and space-and-time-parallel runs (part 1, input and operations).

```
    ...

    <!-- Regex pattern -->
    <patternset name="pattern">
      <pattern name="timing_pat" type="float">
            Time to solution: $jube_pat_fp sec.</pattern>
      <pattern name="niter_pat" type="float">
            Mean number of iterations: $jube_pat_fp</pattern>
    </patternset>

    <!-- Analyze -->
    <analyser name="analyze">
      <use>pattern</use> <!-- use existing patternset -->
      <analyse step="sub_step">
        <file>run.out</file> <!-- file which should be scanned -->
      </analyse>
    </analyser>

    <!-- Create result table -->
    <result>
      <use>analyze</use> <!-- use existing analyser -->
      <table name="result" style="pretty" sort="space_size">
        <column>nnodes</column>
        <column>ntasks</column>
        <column>space_size</column>
        <column>mpi</column>
        <column>timing_pat</column>
        <column>niter_pat</column>
      </table>
    </result>

  </benchmark>
</jube>
```

Listing 2: XML input file for JUBE running space-parallel and space-and-time-parallel runs (part 2, output and analysis).

The runs were set up and executed using JUBE. The corresponding XML file is shown in Listings 1 and 2. The first listing contains the input and operations part of the file and consists of four blocks:

1. the parameter set (lines 6–16),
2. the rules for substituting the parameter values in the template to build the executable (lines 18–27),
3. the list of files to copy over to the run directory (lines 29–33),
4. and the operations part, where the shell command for submitting the job is given (lines 35–42).

While the last two are rather straightforward and do not require too much of the user's attention, the first two are the ones where the simulation and run parameters find their way into the actual execution. In lines 8–12, the number of compute nodes and the number of tasks (or cores) are set up. Using the `python` mode in lines 9 and

11, the variable i from line 8 is taken to step simultaneously through the number of nodes and tasks. Without this, for each number of nodes, all number of tasks would be used in separate runs, i.e. instead of 10 runs, we would end up with 100 runs, most of them irrelevant. Then, in lines 13–14, the simulation parameter `space_size` is defined as being equal to the number of tasks. This specifies the number of cores for spatial parallelization. In line 15, two different MPI versions are requested, where the parameter `mpi` is then handled appropriately in the jobscript. For each combination of these parameters, JUBE creates a separate directory with all necessary files and folders. The template jobscript `run_pySDC_AC.tmpl` is replaced by an actual jobscript `run_pySDC_AC.exe`, see line 21, with all parameters in place. An example of a template jobscript can be found in Listing 3.

The second Listing 2 continues the XML file with the output and analysis blocks. We have:

1. the pattern block (lines 3–9), which will be used to extract data from the output files of the simulation,
2. the analyzer (lines 11–17), which simply applies the pattern to the output file,
3. and the result block (lines 19–30) to create a "pretty" table with the results, based on the parameters and the extracted results.

Using a simple Python script, this table can be read again and processed into Fig. 8. With JUBE, this workflow can be completely automated using only a few configuration files and a post-processing script. All relevant configuration files can be found in the project folder.

4.2 Performance Analysis with Score-P, Scalasca, and Vampir

Performance analysis of a parallel application is not an easy task in general and with non-traditional HPC applications in particular. Python applications are still very rare in the HPC landscape and performance analysis tools (and performance analysts for that matter) are often not yet fully prepared for this scenario. In this section, we present the challenges we faced and the solutions we found to show what tools can do.

We also would like to encourage other application developers to seek assistance from the tool developers and their system administrators when obstacles are encountered in order to get reasonable and satisfactory results.

First measurement attempts The first obstacle we encountered was that the Score-P Python bindings did not build for the tool-chain of Intel compilers and IntelMPI due to an issue with the Intel compiler installation on JURECA. We thus switched to GNU compilers and ParaStationMPI.[3] Using that we were able to obtain a first analysis result.

[3] https://www.par-tec.com/products/parastation-mpi.html.

The workflow to get these results is as follows: After setting up the runs with JUBE XML files as described above, the job is submitted via JUBE using the jobscript generated from the template.

```
1  #!/bin/bash -x
2  #SBATCH --nodes=#NNODES#
3  #SBATCH --ntasks-per-node=#NTASKS#
4  #SBATCH --output=run.out
5  #SBATCH --error=run.err
6  #SBATCH --time=00:05:00
7  #SBATCH --partition=batch
8
9  export MPI=#MPI#
10
11 if [ "$MPI" = "intel" ];
12 ... # logic to distiguish MPI libraries
13 fi
14
15 export SCOREP_EXPERIMENT_DIRECTORY=data/scorep-$MPI
16 export SCOREP_PROFILING_MAX_CALLPATH_DEPTH=90
17 export SCOREP_ENABLE_TRACING=1
18 export SCOREP_METRIC_PAPI=PAPI_TOT_INS
19
20 srun python -m scorep --mpp=mpi run_benchmark.py -n #SPACE_SIZE#
21 srun scout.mpi --time-correct $SCOREP_EXPERIMENT_DIRECTORY/traces.otf2
22 touch ready
```

Listing 3: Jobscript template to run the simulation with profiling and tracing enabled.

Listing 3 shows such a template, where all variables of the form `#NAME#` will be replaced by actual values for the specific run. Lines 2–7 provide the allocation and job information for the Slurm Workload Manager. In lines 9–13, the distinction between different MPI libraries is implemented, using different modules and virtual Python environments (not shown here). Lines 15–18 define flags for the Score-P infrastructure, e.g. tracing is enabled (line 17). Then, line 20 contains the run command, where the Score-P infrastructure is passed using the `-m` switch. This generates both a profile report (profiling is enabled by default) for an analysis with CUBE and OTF2 trace files, which can be analyzed manually with Vampir or automatically with Scalasca. The Scalasca trace analyzer is called on line 21. As `pySDC` is a pure MPI application, `scout.mpi` is used here (there is also a `scout.omp` for OpenMP and a `scout.hyb` for hybrid programs). Note that tracing is enabled manually here, but could be part of the parameter input as described in Sect. 3.5. Finally, line 22 marks this particular run as completed for JUBE. The resulting files can then be read by tools like Vampir and CUBE.

Using this setup, we were able to get a first usable measurement. We used filtering and Score-P's manual instrumentation API to mark the interesting parts of the application. In Listing 4, a mock-up of a PFASST implementation is shown. Here, after importing the Python module `scorep.user`, separate regions can be defined using

```
1  from mpi4py import MPI
2  from pySDC.core.Controller import controller
3
4  import scorep.user as spu
5
6  ...
7
8  def run_pfasst(*args, **kwargs):
9      ...
10
11     while not done:
12         ...
13         name = f'REGION -- IT_FINE -- {my_rank}'
14         spu.region_begin(name)
15         controller.do_fine_sweep()
16         spu.region_end(name)
17         ...
18         name = f'REGION -- IT_DOWN -- {my_rank}'
19         spu.region_begin(name)
20         controller.transfer_down()
21         spu.region_end(name)
22         ...
23         name = f'REGION -- IT_COARSE -- {my_rank}'
24         spu.region_begin(name)
25         controller.do_coarse_sweep()
26         spu.region_end(name)
27         ...
28         name = f'REGION -- IT_UP -- {my_rank}'
29         spu.region_begin(name)
30         controller.transfer_up()
31         spu.region_end(name)
32         ...
33         name = f'REGION -- IT_CHECK -- {my_rank}'
34         spu.region_begin(name)
35         controller.check_convergence()
36         spu.region_end(name)
37         ...
38
39     ...
40
41 ...
```

Listing 4: Pseudo code of a PFASST implementation using Score-P regions

`region_start` and `region_end`, see e.g. lines 14 and 16. This information will then be available, e.g. for filtering in Vampir.

Analysis then showed that the algorithm outlined in Fig. 1b worked as expected, at least in principle. This can be seen in Fig. 9: the bottom part shows exactly a transposed version of the original communication and workflow structure as expected

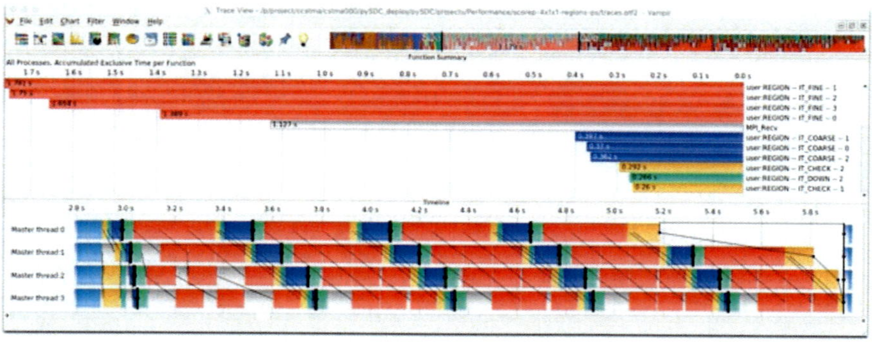

Fig. 9 Vampir visualization: user-defined regions, only a single iteration (ParaStation MPI, 4 processes in time, 1 in space)

from Fig. 1b. The middle part shows the amount of time spend in the different regions: the vast majority of the computation time (70%) is spent in the fine sweep, and only about 3% in the coarse sweep.

Another, more high-level overview of the parallel performance can be gained with the Advisor plugin of Cube [23]. This prints the efficiency metrics developed in the POP project[4] for the entire execution or an arbitrary phase of the application. Figure 10a shows a screenshot of the Advisor result for the computational part of pySDC, i.e. omitting initialization and finalization.

The main value to look for is "Parallel Efficiency", which reveals the inefficiency in splitting computation over processes and then communicating data between processes. In this case, the "Parallel Efficiency", which is defined as the product of "Load Balance" and "Communication Efficiency", is 79%, which is worse than what we expected for this small test case. We know from Sect. 2.2 that due to the sequential coarse level and the predictor, PFASST runs will always show slight load imbalances, so the "Load Balance" value of 89% is understood.

However, the "Communication Efficiency" of 88% is way below our expectations. A "Serialisation Efficiency" of 98% indicates that there is hardly any waiting time. The "Transfer Efficiency" of 90% means we lose significant time due to data transfers. This was not expected so we assumed either an issue with the implementation of the algorithm or the system environment. A Scalasca analysis showed that the slight serialization inefficiency originates from a "Late Receiver pattern" (see Fig. 5) in the fine sweep phase and a "Late Broadcast" after each time step, but did not reveal the reason for the loss in transfer efficiency. A closer look with Vampir at just a single iteration, as shown in Fig. 9, finally reveals the issue.

The implementation of pySDC uses non-blocking MPI communication in order to overlap computation and communication. However, Fig. 9 clearly shows that this does not work as expected.

In the time of the analysis of the ParaStationMPI runs there was an update of the JURECA software environment which finally enabled the support of the Score-P

[4] https://pop-coe.eu/node/69.

Python wrappers for the Intel compilers and IntelMPI. So we performed the same analysis again for this constellation, the one we originally intended to analyze anyway. Surprisingly, the results looked much better now. The Cube Advisor analysis now showed nearly perfect Transfer Efficiency and, subsequently, a much improved Parallel Efficiency, see Fig. 10b.

Vampir further confirms a very good overlap of computation and communication, the way the implementation intended it to be, see Fig. 11.

Eye for the detail Thus, the natural question to ask is where these differences between the exact same code running on two different toolchains come from. Further investigation showed that the installation of ParaStationMPI provided on JURECA does not provide an MPI progress thread, i.e. MPI communication cannot be performed asynchronously and thus overlapping computation and communication is not possible. IntelMPI on the other hand always uses a progress thread if not explicitly disabled via an environment variable. With a newly installed test version of ParaStationMPI, where an MPI progress thread has been enabled, the overlap of computation and communication is possible there, too. We then see a similar performance of `pySDC` using the new ParaStationMPI and IntelMPI.

Even though the overlap problem does not seem to be that much of an issue for this small test case, where just 8% efficiency could be gained, we want to emphasize that these little issues can become severe ones when scaling up. Figure 12 shows the average time per call of the fine sweep, as calculated by CUBE. In the Intel case with overlap, we see that the fine sweep time is very balanced across the processes (Fig. 12b). In the ParaStationMPI case, we see that the fine sweep time increases with the process number (Fig. 12a). This problem will likely become worse when the problem size is increased, thus limiting the maximum number of processes that can be utilized.

The scaling tests as well as the performance analysis made for this work are rather small compared to what joined space and time parallelism can do. The difference when using space-parallel solvers can be quite substantial for the analysis ranging from larger datasets for the analysis and visualization to more complex communication patterns. In addition, the issues experienced can differ, as we already see for the test case at hand. In Fig. 13, we now use two processes in space and four in time. There is still unnecessary waiting time, but its impact is much smaller. This is due to the fact that progress of MPI calls does not depend on the MPI communicator, i.e. for each application of the space-parallel FFT solver progress does happen even in the time-communicator. A more thorough and in-depth analysis of large-scale runs is currently underway together with the POP CoE and we will report on the outcome of this in a future publication.

5 Conclusion and Outlook

In this paper, we performed and analyzed parallel runs of the PFASST implementation `pySDC` using the performance tools Score-P, CUBE, Scalasca, and Vampir as well

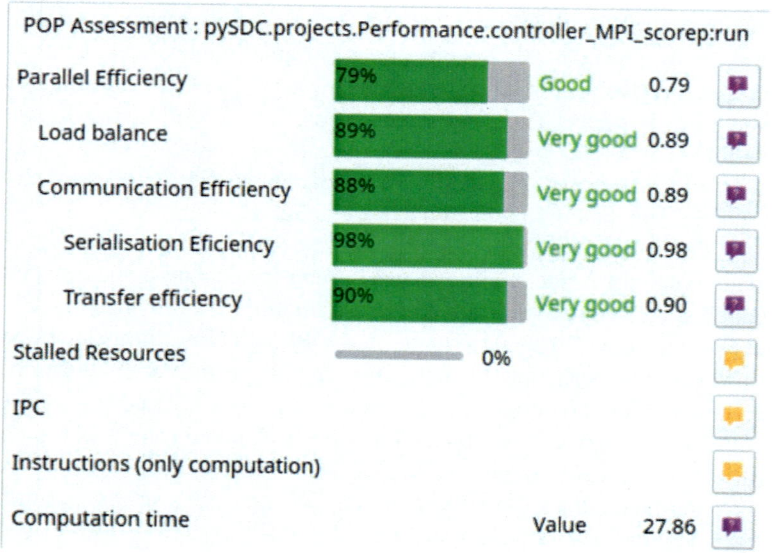

(a) Cube Advisor showing the POP metrics for pySDC with ParaStationMPI.

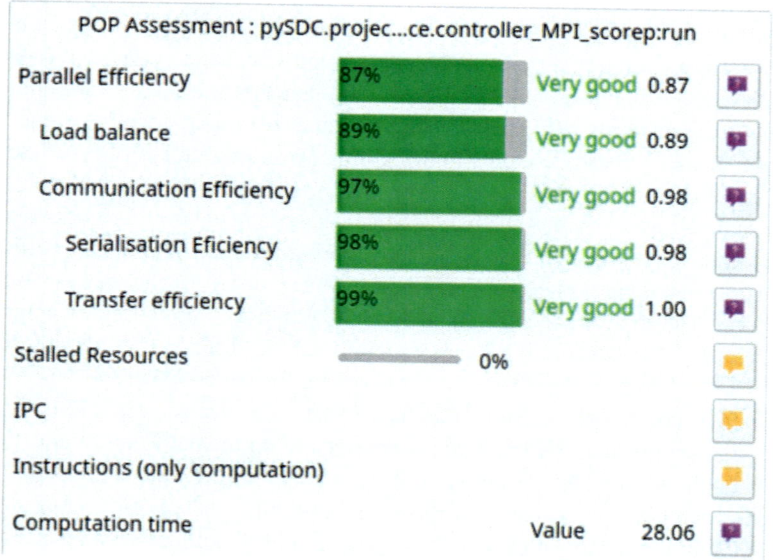

(b) Cube Advisor showing the POP metrics for pySDC with IntelMPI.

Fig. 10 Cube Advisor showing the POP metrics for pySDC

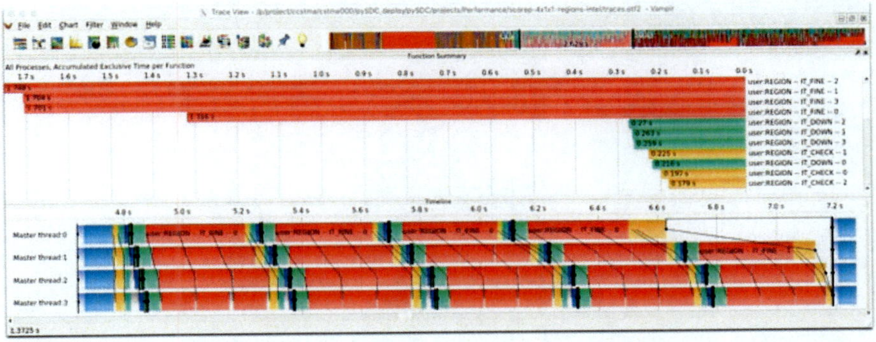

Fig. 11 Vampir visualization: user-defined regions, only a single iteration (Intel MPI, 4 processes in time, 1 in space)

as the benchmarking environment JUBE. While the implementation complexity of a time-parallel method may vary, with standard Parareal being on one side of the spectrum and methods like PFASST on the other, it is crucial to check and analyze the actual performance of the code. This is particularly true for most time-parallel methods with their theoretically grounded low parallel efficiency since here problems in the communication can easily be mistaken for method-related shortcomings.

As we have shown, the performance analysis tools in the Score-P ecosystem cannot only be used to identify tuning potential but also allow to easily check for bugs and unexpected behavior, without the need to do "print"-debugging. While methods like Parareal may be straightforward to implement, PFASST is not, in particular, due to many edge cases which the code needs to take care of. For example, in the standard PFASST implementation, the residual is checked locally for each time step individually so that a process working on a later time step could, erroneously, decide to stop although the iterations on previous time steps still run. Vice versa, when previous time steps did converge, the processes dealing with later ones should not expect to receive new data. Depending on the implementation, those cases could lead to deadlocks (the "good" case) or to unexpected results (the "bad" case), e.g. when one-sided communication is used, or other unwanted behavior. Many of these issues can be checked by looking at the gathered data after an instrumented run. This does not, however, replace a careful design of the code, testing, benchmarking, verification, and, sometimes, rethinking.

We saw for `pySDC` that already the choice of the MPI implementation can influence the performance quite severely, let alone the unexpected deviation from the intended workflow of the method. Performance tools as the ones presented here help to verify (or falsify) that the implementation of an algorithm actually does what the developers think it does. While there is a lot of documentation on these tools available, it is extremely helpful and productive to get in touch with the core developers, either directly or by attending one of the tutorials, e.g. provided by the VI-HPS

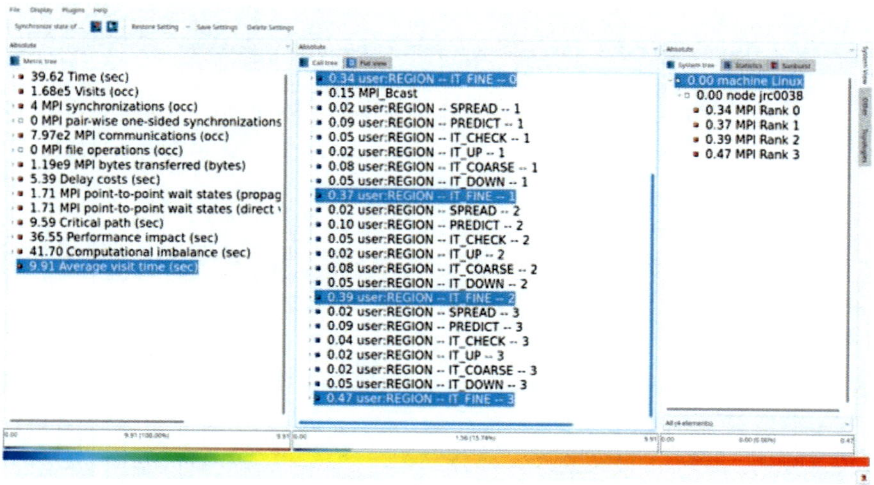

(a) Cube screenshot showing the average time per call of the fine sweep for ParaStationMPI. Time increases with process number.

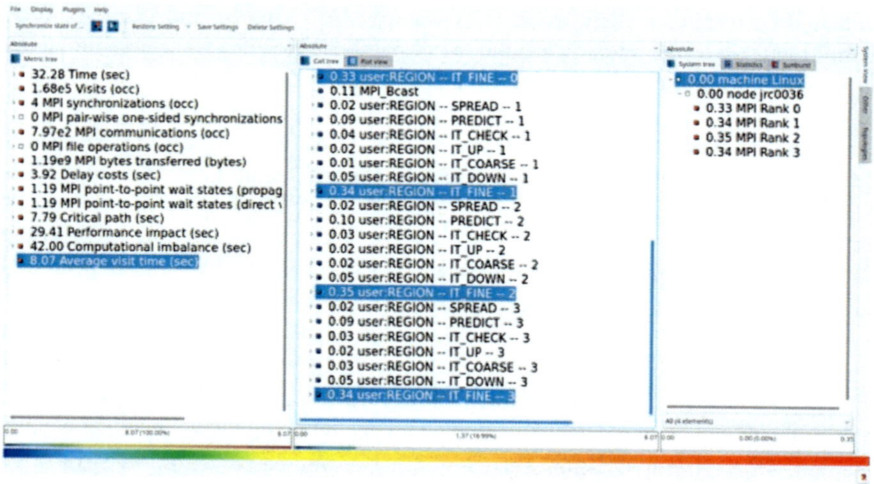

(b) Cube screenshot showing the average time per call of the fine sweep for IntelMPI. Time is well balanced across the processes.

Fig. 12 Cube screenshots showing the average time per call of the fine sweep

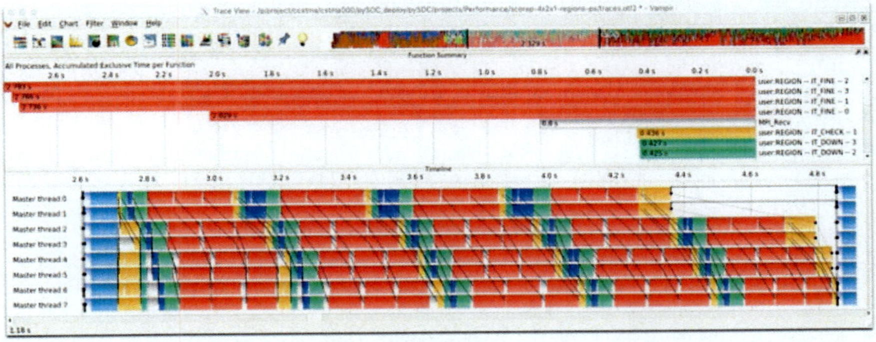

Fig. 13 Vampir visualization: user-defined regions, only a single iteration (ParaStation MPI, 4 processes in time, 2 in space)

through the Tuning Workshop series.[5] This way, many of the pitfalls and sources of frustration can be avoided and the full potential of these tools becomes visible.

In order to set up experiments using parallel codes in a structured way, be it for performance analysis, parameter studies, or scaling tests, tools like JUBE can be used to ease the management of submission, monitoring, and post-processing of the jobs. Besides parameters for the model, the methods in space and time, the iteration, and so on, the application of time-parallel methods in combination with spatial parallelism adds another level of complexity, which becomes manageable with tools like JUBE.

Acknowledgements Parts of this work have received funding from the European Union's Horizon 2020 research and innovation programme under grant agreements No 676553 and 824080. RS thankfully acknowledges the financial support by the German Federal Ministry of Education and Research through the ParaPhase project within the framework "IKT 2020 - Forschung für Innovationen" (project number 01IH15005A).

References

1. Adhianto, L., Banerjee, S., Fagan, M., Krentel, M., Marin, G., Mellor-Crummey, J., Tallent, N.R.: HPCToolkit: Tools for performance analysis of optimized parallel programs. Concurrency and Computation: Practice and Experience **22**(6), 685–701 (2010)
2. Böhme, D., Wolf, F., de Supinski, B.R., Schulz, M., Geimer, M.: Scalable critical-path based performance analysis. In: 2012 IEEE 26th International Parallel and Distributed Processing Symposium, pp. 1330–1340. IEEE (2012)
3. Bolten, M., Moser, D., Speck, R.: A multigrid perspective on the parallel full approximation scheme in space and time. Numerical Linear Algebra with Applications **24**(6), e2110–n/a (2017). DOI https://doi.org/10.1002/nla.2110. E2110 nla.2110
4. Bolten, M., Moser, D., Speck, R.: Asymptotic convergence of the parallel full approximation scheme in space and time for linear problems. Numerical linear algebra with applications **25**(6),

[5] https://www.vi-hps.org/training/tws/tuning-workshop-series.html.

e2208 – (2018). DOI https://doi.org/10.1002/nla.2208. URL https://juser.fz-juelich.de/record/857114
5. Bradley, T.: GPU Performance Analysis and Optimisation. NVIDIA Corporation (2012)
6. Center, B.S.: Website for POP CoE (2019). URL https://pop-coe.eu/. [Online; accessed August 13, 2019]
7. Dalcin, L.D., Paz, R.R., Kler, P.A., Cosimo, A.: Parallel distributed computing using python. Advances in Water Resources **34**(9), 1124–1139 (2011). DOI https://doi.org/10.1016/j.advwatres.2011.04.013. URL http://www.sciencedirect.com/science/article/pii/S0309170811000777. New Computational Methods and Software Tools
8. Dalcin, Lisandro and Mortensen, Mikael and Keyes, David E: Fast parallel multidimensional FFT using advanced MPI. Journal of Parallel and Distributed Computing (2019). DOI https://doi.org/10.1016/j.jpdc.2019.02.006
9. Dutt, A., Greengard, L., Rokhlin, V.: Spectral deferred correction methods for ordinary differential equations. BIT Numerical Mathematics **40**(2), 241–266 (2000). DOI https://doi.org/10.1023/A:1022338906936
10. Emmett, M., Minion, M.L.: Toward an efficient parallel in time method for partial differential equations. Communications in Applied Mathematics and Computational Science **7**, 105–132 (2012). DOI https://doi.org/10.2140/camcos.2012.7.105
11. Emmett, M., Minion, M.L.: Efficient implementation of a multi-level parallel in time algorithm. In: Domain Decomposition Methods in Science and Engineering XXI, *Lecture Notes in Computational Science and Engineering*, vol. 98, pp. 359–366. Springer International Publishing (2014). DOI https://doi.org/10.1007/978-3-319-05789-7_33
12. Eschweiler, D., Wagner, M., Geimer, M., Knüpfer, A., Nagel, W.E., Wolf, F.: Open Trace Format 2 - The next generation of scalable trace formats and support libraries. In: Proc. of the Intl. Conference on Parallel Computing (ParCo), Ghent, Belgium, August 30–September 2 2011, *Advances in Parallel Computing*, vol. 22, pp. 481–490. IOS Press (2012). DOI https://doi.org/10.3233/978-1-61499-041-3-481
13. Feld, C., Convent, S., Hermanns, M.A., Protze, J., Geimer, M., Mohr, B.: Score-p and ompt: Navigating the perils of callback-driven parallel runtime introspection. In: X. Fan, B.R. de Supinski, O. Sinnen, N. Giacaman (eds.) OpenMP: Conquering the Full Hardware Spectrum, pp. 21–35. Springer International Publishing, Cham (2019)
14. Gander, M.J.: 50 years of Time Parallel Time Integration. In: Multiple Shooting and Time Domain Decomposition. Springer (2015). DOI https://doi.org/10.1007/978-3-319-23321-5_3
15. Geimer, M., Wolf, F., Wylie, B.J.N., Ábrahám, E., Becker, D., Mohr, B.: The SCALASCA performance toolset architecture. In: International Workshop on Scalable Tools for High-End Computing (STHEC), Kos, Greece, pp. 51–65 (2008)
16. Geimer, M., Saviankou, P., Strube, A., Szebenyi, Z., Wolf, F., Wylie, B.J.N.: Further improving the scalability of the scalasca toolset. In: K. Jónasson (ed.) Applied Parallel and Scientific Computing, pp. 463–473. Springer Berlin Heidelberg, Berlin, Heidelberg (2012)
17. Gocht, A., Schöne, R., Frenzel, J.: Advanced Python Performance Monitoring with Score-P. In: Tools for High Performance Computing 2019, p. to appear. Springer International Publishing (2019)
18. Harlacher, M., Calotoiu, A., Dennis, J., Wolf, F.: Analysing the Scalability of Climate Codes Using New Features of Scalasca. In: K. Binder, M. Müller, M. Kremer, A. Schnurpfeil (eds.) Proc. of the John von Neumann Institute for Computing (NIC) Symposium 2016, Juelich, Germany, *NIC Series*, vol. 48, pp. 343–352. Forschungszentrum Jülich, John von Neumann-Institut for Computing (2016)
19. Hermanns, M.A., Geimer, M., Mohr, B., Wolf, F.: Trace-based detection of lock contention in MPI one-sided communication. In: C. Niethammer, J. Gracia, T. Hilbrich, A. Knüpfer, M.M. Resch, W.E. Nagel (eds.) Tools for High Performance Computing 2016, Proc. of the 10th Parallel Tools Workshop, Stuttgart, Germany, October 2016, pp. 97–114. Springer (2017). DOI https://doi.org/10.1007/978-3-319-56702-0_6. URL http://juser.fz-juelich.de/record/830159
20. Huang, J., Jia, J., Minion, M.: Accelerating the convergence of spectral deferred correction methods. Journal of Computational Physics **214**(2), 633–656 (2006)

21. Jülich Supercomputing Centre: JURECA: General-purpose supercomputer at Jülich Supercomputing Centre. Journal of large-scale research facilities **2**(A62) (2016). DOI https://doi.org/10.17815/jlsrf-2-121
22. Knobloch, M., Mohr, B.: Tools for GPU Computing – Debugging and Performance Analysis of Heterogenous HPC Applications. Supercomputing Frontiers and Innovations **7**(1) (2020). URL https://superfri.org/superfri/article/view/311
23. Knobloch, M., Saviankou, P., Schlütter, M., Visser, A., Mohr, B.: A picture is worth a thousand numbers – Enhancing Cube's analysis capabilities with plugins. In: Tools for High Performance Computing 2019 (tbp)
24. Knüpfer, A., Brunst, H., Doleschal, J., Jurenz, M., Lieber, M., Mickler, H., Müller, M.S., Nagel, W.E.: The Vampir Performance Analysis Tool-Set. In: M. Resch, R. Keller, V. Himmler, B. Krammer, A. Schulz (eds.) Tools for High Performance Computing, pp. 139–155. Springer Berlin / Heidelberg (2008). DOI https://doi.org/10.1007/978-3-540-68564-7_9
25. Knüpfer, A., Rössel, C., an Mey, D., Biersdorff, S., Diethelm, K., Eschweiler, D., Geimer, M., Gerndt, M., Lorenz, D., Malony, A.D., Nagel, W.E., Oleynik, Y., Philippen, P., Saviankou, P., Schmidl, D., Shende, S.S., Tschüter, R., Wagner, M., Wesarg, B., Wolf, F.: Score-P – A joint performance measurement run-time infrastructure for Periscope, Scalasca, TAU, and Vampir. In: Proc. of the 5th Int'l Workshop on Parallel Tools for High Performance Computing, September 2011, Dresden, pp. 79–91. Springer (2012). DOI https://doi.org/10.1007/978-3-642-31476-6_7
26. LLNL: Website for `XBraid` (2018). URL https://www.llnl.gov/casc/xbraid. [Online; accessed July 30, 2018]
27. Lührs, S., Rohe, D., Schnurpfeil, A., Thust, K., Frings, W.: Flexible and Generic Workflow Management. In: Parallel Computing: On the Road to Exascale, *Advances in parallel computing*, vol. 27, pp. 431–438. International Conference on Parallel Computing 2015, Edinburgh (United Kingdom), 1 Sep 2015–4 Sep 2015, IOS Press, Amsterdam (2016). DOI https://doi.org/10.3233/978-1-61499-621-7-431. URL http://juser.fz-juelich.de/record/808798
28. Minion, M., Emmett, M.: Website for `libpfasst` (2019). URL https://github.com/libpfasst/LibPFASST. [Online; accessed August 13, 2019]
29. Minion, M.L.: A hybrid parareal spectral deferred corrections method. Communications in Applied Mathematics and Computational Science **5**(2), 265–301 (2010). DOI https://doi.org/10.2140/camcos.2010.5.265
30. Mix, H., Herold, C., Weber, M.: Visualization of Multi-layer I/O Performance in Vampir. In: Parallel and Distributed Processing Symposium Workshop (IPDPSW), 2018 IEEE International (2018)
31. Ong, B.W., Schroder, J.B.: Applications of time parallelization. Computing and Visualization in Science **23**(1), 1–15 (2020)
32. Ong, B.W., Haynes, R.D., Ladd, K.: Algorithm 965: RIDC Methods: A Family of Parallel Time Integrators. ACM Trans. Math. Softw. **43**(1), 8:1–8:13 (2016). DOI https://doi.org/10.1145/2964377
33. Pillet, V., Labarta, J., Cortes, T., Girona, S.: Paraver: A tool to visualize and analyze parallel code. In: Proceedings of WoTUG-18: transputer and occam developments, vol. 44, pp. 17–31. Citeseer (1995)
34. Reinders, J.: Vtune performance analyzer essentials. Intel Press (2005)
35. Ruprecht, D., Speck, R.: Spectral deferred corrections with fast-wave slow-wave splitting. SIAM Journal on Scientific Computing **38**(4), A2535–A2557 (2016)
36. Saviankou, P., Knobloch, M., Visser, A., Mohr, B.: Cube v4: From performance report explorer to performance analysis tool. In: Proceedings of the International Conference on Computational Science, ICCS 2015, Computational Science at the Gates of Nature, Reykjavík, Iceland, 1–3 June, 2015, pp. 1343–1352 (2015). DOI https://doi.org/10.1016/j.procs.2015.05.320
37. Saviankou, P., Knobloch, M., Visser, A., Mohr, B.: Cube v4: From performance report explorer to performance analysis tool. Procedia Computer Science **51**, 1343–1352 (2015)
38. Sharples, W., Zhukov, I., Geimer, M., Goergen, K., Luehrs, S., Breuer, T., Naz, B., Kulkarni, K., Brdar, S., Kollet, S.: A run control framework to streamline profiling, porting, and tuning

simulation runs and provenance tracking of geoscientific applications. Geoscientific Model Development **11**(7), 2875–2895 (2018). DOI https://doi.org/10.5194/gmd-11-2875-2018
39. Shende, S.S., Malony, A.D.: The TAU parallel performance system. The International Journal of High Performance Computing Applications **20**(2), 287–311 (2006)
40. Speck, R.: Algorithm 997: pySDC - Prototyping Spectral Deferred Corrections. ACM Transactions on Mathematical Software **45**(3) (2019). DOI https://doi.org/10.1145/3310410
41. Speck, R.: Parallel-in-time/pysdc: The performance release (2019). DOI https://doi.org/10.5281/zenodo.3407254
42. Speck, R.: Website for `pySDC` (2019). URL https://parallel-in-time.org/pySDC/. [Online; accessed August 13, 2019]
43. Terpstra, D., Jagode, H., You, H., Dongarra, J.: Collecting performance data with papi-c. In: Tools for High Performance Computing 2009, pp. 157–173. Springer (2010)
44. Treibig, J., Hager, G., Wellein, G.: Likwid: A lightweight performance-oriented tool suite for x86 multicore environments. In: 2010 39th International Conference on Parallel Processing Workshops, pp. 207–216. IEEE (2010)
45. Weiser, M.: Faster SDC convergence on non-equidistant grids by DIRK sweeps. BIT Numerical Mathematics **55**(4), 1219–1241 (2014)
46. Zhang, J., Du, Q.: Numerical Studies of Discrete Approximations to the Allen-Cahn Equation in the Sharp Interface Limit. SIAM Journal on Scientific Computing **31**(4), 3042–3063 (2009). DOI https://doi.org/10.1137/080738398
47. Zhukov, I., Feld, C., Geimer, M., Knobloch, M., Mohr, B., Saviankou, P.: Scalasca v2: Back to the future. In: Proc. of Tools for High Performance Computing 2014, pp. 1–24. Springer (2015). DOI https://doi.org/10.1007/978-3-319-16012-2_1

Twelve Ways to Fool the Masses When Giving Parallel-in-Time Results

Sebastian Götschel, Michael Minion, Daniel Ruprecht, and Robert Speck

Abstract Getting good speedup—let alone high parallel efficiency—for parallel-in-time (PinT) integration examples can be frustratingly difficult. The high complexity and large number of parameters in PinT methods can easily (and unintentionally) lead to numerical experiments that overestimate the algorithm's performance. In the tradition of Bailey's article "Twelve ways to fool the masses when giving performance results on parallel computers", we discuss and demonstrate pitfalls to avoid when evaluating the performance of PinT methods. Despite being written in a light-hearted tone, this paper is intended to raise awareness that there are many ways to unintentionally fool yourself and others and that by avoiding these fallacies more meaningful PinT performance results can be obtained.

1 Introduction

The trend toward extreme parallelism in high-performance computing requires novel numerical algorithms to translate the raw computing power of hardware into application performance [5]. Methods for the approximation of time-dependent partial differential equations, which are used in models in a very wide range of disciplines

S. Götschel · D. Ruprecht
Chair Computational Mathematics, Institute of Mathematics, Hamburg University of Technology, Am Schwarzenberg-Campus 3, 21073 Hamburg, Germany
e-mail: sebastian.goetschel@tuhh.de

D. Ruprecht
e-mail: ruprecht@tuhh.de

M. Minion (✉)
Lawrence Berkeley National Laboratory, 1 Cyclotron Road, Berkeley, CA 94720, USA
e-mail: mlminion@lbl.gov

R. Speck
Jülich Supercomputing Centre, Forschungszentrum Jülich GmbH, 52425 Jülich, Germany
e-mail: r.speck@fz-juelich.de

© The Author(s), under exclusive license to Springer Nature Switzerland AG 2021

B. Ong et al. (eds.), *Parallel-in-Time Integration Methods*, Springer Proceedings in Mathematics & Statistics 356, https://doi.org/10.1007/978-3-030-75933-9_4

from engineering to physics, biology, or even sociology, pose a particular challenge in this respect. Parallelization of algorithms discretizing the spatial dimension via a form of domain decomposition is quite natural and has been an active research topic for decades. Exploiting parallelization in the time direction is less intuitive as time has a clear direction of information transport. Traditional algorithms for temporal integration employ a step-by-step procedure that is difficult to parallelize. In many applications, this sequential treatment of temporal integration has become a bottleneck in massively parallel simulations.

Parallel-in-time (PinT) methods, i.e. methods that offer at least some degree of concurrency, are advertised as a possible solution to this temporal bottleneck. The concept was pioneered by Nievergelt in 1964 [15] but has only really gained traction in the last two decades [7]. By now, the effectiveness of PinT has been well established for examples ranging from the linear heat equation in one dimension to more complex highly diffusive problems in more than one dimension. More importantly, there is now ample evidence that different PinT methods can deliver measurable reduction in solution times on real-life HPC systems for a wide variety of problems. Ong and Schroder [16] and Gander [7] provide overviews of the literature, and a good resource for further reading is also given by the community website https://parallel-in-time.org/.

PinT methods differ from space-parallel algorithms or parallel methods for operations like the FFT in that they do not simply parallelize a serial algorithm to reduce its run time.[1] Instead, serial time-stepping is usually replaced with a computationally more costly and typically iterative procedure that is amenable to parallelization. Such a procedure will run much slower in serial but can overtake serial time-stepping in speed if sufficiently many processors are employed. This makes a fair assessment of performance much harder since there is no clear baseline to compare against. Together with a large number of parameters and inherent complexities in PinT methods and PDEs themselves, there are thus many sometimes subtle ways to fool oneself (and the masses) when assessing performance. We will demonstrate various ways to produce results that seem to demonstrate speedup but are essentially meaningless. The paper is written in a similar spirit as other "ways to fool the masses" papers first introduced in [3] that inspired a series of similarly helpful papers in related areas [4, 9–11, 14, 17, 18]. One departure from the canon here is that we provide actual examples to demonstrate the *Ways* as we present them. Despite the light-hearted, sometimes even sarcastic tone of the discussion, the numerical examples are similar to experiments one could do for evaluating the performance of PinT methods.

Some of the *Ways* we present are specific to PinT while others, although formulated in "PinT language" correspond to broader concepts from parallel computing. This illustrates another important fact about PinT: while the algorithms often dig deeply into the applied mathematics toolkit, their relevance is exclusively due to the architectural specifics of modern high-performance computing systems. This inherent cross-disciplinarity is another complicating factor when trying to do fair perfor-

[1] One exception are so-called "parallel-across-the-method" PinT methods in the terminology by Gear [8] that can deliver smaller scale parallelism.

mance assessments. Lastly, we note that this paper was first presented in a shorter form as a conference talk at the 9th Parallel-in-Time Workshop held (virtually) in June 2020. Hence, some of the *Ways* are more relevant to live presentations, although all should be considered in both written and live scenarios. In the next section, we present the 12 *Ways* with a series of numerical examples before concluding with some more serious comments in Sect. 3.

2 Fool the Masses (and Yourself)

2.1 *Choose Your Problem Wisely!*

If you really want to impress the masses with your PinT results, you will want to show as big a parallel speedup as possible, hence you will want to use a lot of processors in the time direction. If you are using, for example, Parareal [13], a theoretical model for speedup is given by the expression

$$S_{\text{theory}} = \frac{N_P}{N_P\alpha + K(1+\alpha)}, \tag{1}$$

where N_P is the number of processors, α is the ratio of the cost of the coarse propagator $\mathcal{G}$ compared to the fine propagator $\mathcal{F}$, and K is the total number of iterations needed for convergence. Hence to get a large speedup that will impress the masses, we need to choose N_P to be large, α to be small, and hope K is small as well. A common choice for parareal is to have $\mathcal{G}$ be one step of some method and $\mathcal{F}$ be $N_{\mathcal{F}}$ steps of the same method so that $\alpha = 1/N_{\mathcal{F}}$ is small. But note that this already means that the total number of time steps corresponding to the serial method is now $N_P N_{\mathcal{F}}$. Hence, we want to choose an equation and problem parameters for which very many time steps can be employed, while still showing good speedup without raising any suspicions that the problem is too "easy". The first example suggests some *Ways* to pull off this perilous balancing act.

In this example, we use the following nonlinear advection-diffusion-reaction equation:

$$u_t = \upsilon u_x + \gamma u u_x + \nu u_{xx} + \beta u(a-u)(b-u),$$

where the constants υ, γ, ν, β, a, and b determine the strength of each term. In order to squeeze in the massive number of time steps we need for good speedup, we choose a long time interval over which to integrate, $t \in [0, T_F]$, with $T_F = 30$. The initial condition is given on the interval $[0, 2\pi]$ by

$$u(x, 0) = 1 - d(1 - e^{-(x-\pi)^4/\sigma}).$$

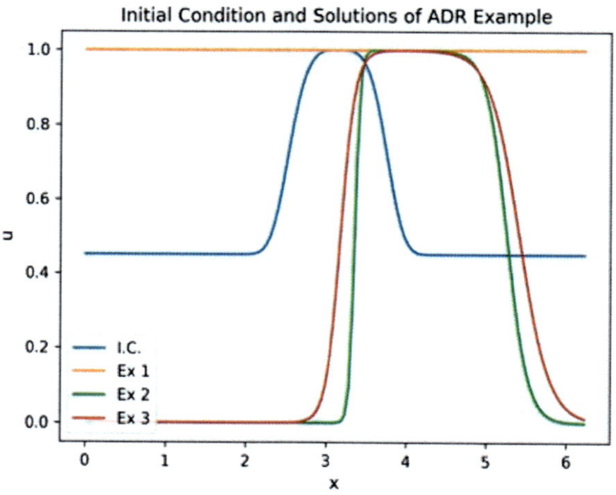

Fig. 1 Initial solution and solution at $t = 30$ for the advection-diffusion-reaction problem demonstrating the significant effect that parameter selection can have on the dynamics and subsequent PinT speedup discussed in *Ways* 1–4

If you are presenting this example in front of an audience, try to get all the equations with parameters on one slide and then move on before defining them.

> **Way 1.** Choose a seemingly complicated problem with lots of parameters which you define later (or not at all).

For the first numerical test, we choose $N_P = 200$ processors and use a fourth-order IMEX Runge-Kutta method and a pseudospectral discretization in space using 128 grid points, where the linear advection and diffusion terms are treated implicitly. We use one time step for $\mathcal{G}$ and $N_{\mathcal{F}} = 64$ steps for $\mathcal{F}$. Since the method is spectrally accurate in space, it gives us cover to use a lot of time steps (more on that later). We set the stopping criterion for Parareal to be when the increment in the iteration is below 10^{-9}, and $v = -0.5, \gamma = 0.25, \nu = 0.01, \beta = -5, a = 1, b = 0$, and $d = 0.55$ (see also Appendix 1). For these values, Parareal converges on the entire time interval in three iterations. The theoretical speedup given by Eq. (1) is 32.4. Not bad!

If we explore no further, we might have fooled the masses. How did we manage? Consider a plot of the initial condition and solution at the final time for this problem shown in Fig. 1, with the blue and orange lines, respectively. The lesson here is

> **Way 2.** Quietly use an initial condition and/or problem parameters for which the solution tends to a steady state. But do not show the actual solution.

If we repeat this experiment changing only one parameter, $b = 0.5$, the number of Parareal iterations needed for convergence jumps to $K = 10$ for a less impressive theoretical speedup of 15.05. In this case, the solution quickly evolves not to constant state, but a steady bump moving at constant speed (the green line in Fig. 1). This raises another important point to fool the masses:

> **Way 3.** Do not show the sensitivity of your results to problem parameter changes. Find the best one and let the audience think the behavior is generic.

Sometimes you might be faced with a situation like the second case above and not know how to get a better speedup. One suggestion is to add more diffusion. Using the same parameters except increasing the diffusive coefficient to $\nu = 0.04$ reduces the number of iterations for convergence to $K = 5$ with a theoretical speedup of 24.38. The solution of this third example is shown by the red line Fig. 1. If you can't add diffusion directly, using a diffusive discretization for advection like first-order upwind finite differences can sometimes do the trick while avoiding the need to explicitly admit to the audience that you needed to increase the amount of "diffusion".

Way 4. If you are not completely thrilled about the speedup because the number of iterations K is too high, try adding more diffusion. You might have to experiment a little to find just the right amount.

2.2 *Over-Resolve the Solution! Then Over-Resolve Some More*

After carefully choosing your test problem, there are ample additional opportunities to boost the parallel performance of your numerical results. The next set of *Ways* considers the effect of spatial and temporal resolution. We consider the 1D nonlinear Schrödinger equation

$$u_t = \mathrm{i}\nabla^2 u + 2\mathrm{i}\,|u|^2\,u \tag{2}$$

with periodic boundary conditions in $[0, 2\pi]$ and the exact solution as given by Aktosun et al. [1], which we also use for the initial condition at time $t_0 = 0$.

This is a notoriously hard problem for out-of-the-box PinT methods, but we are optimistic and give Parareal a try. We use a second-order IMEX Runge-Kutta method by Ascher et al. [2] with $N_{\mathcal{F}} = 1\,024$ steps and $N_{\mathcal{G}} = 32$ coarse steps for each of the 32 processors. In space, we again use a pseudospectral method with the linear part treated implicitly and $N_x = 32$ degrees-of-freedom. The estimated speedup can be found in Fig. 2a. Using $K = 5$ iterations, we obtain a solution about 6.24 times faster when running with 32 instead of 1 processor. All runs achieve the same accuracy of 5.8×10^{-5} and it looks like speedup in time can be easily achieved after all.

Yet, although the accuracy compared to the exact solution is the same for all runs, the temporal resolution is way too high for this problem, masking the effect of coarsening in time. The spatial error is dominating and instead of $32 \times 1\,024 = 32\,768$ overall time steps, only $32 \times 32 = 1\,024$ are actually needed to balance spatial and temporal error. Therefore, the coarse level already solves the problem quite well: speedup only comes from over-resolving in time.

If we instead choose $N_{\mathcal{F}} = 32$ time steps on the fine level instead and the same coarsening factor of 32 ($N_{\mathcal{G}} = 1$), we get no speedup at all—see the red curve/diamond markers in Fig. 2b. Using a less drastic coarsening factor of 4 leads to a maximum speedup of 1.78 with 32 processors (blue curve/square markers), which

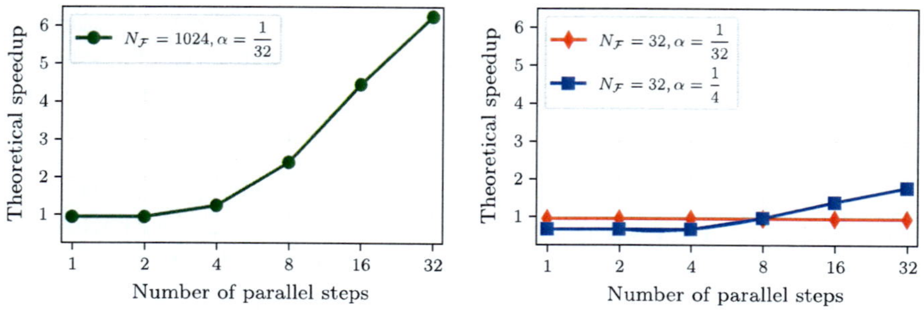

(a) Deceivingly good speedup (b) Not so good speedup

Fig. 2 Estimated speedup for Parareal runs of the nonlinear Schrödinger example (2) demonstrating the effect of over-resolution in time (*Way* 5)

is underwhelming and frustrating and not what we would prefer to present in public. Lesson learned:

> **Way 5.** Make Δt so small that the coarse integrator is already accurate. Never check if a larger Δt might give you the same solution.

The astute readers may have noticed we also used this trick to a lesser extent in the advection-diffusion-reaction example above.

A similar effect can be achieved when considering coarsening in space. Since we have learned that more parameters are always good to fool the masses, we now use PFASST [6] instead of Parareal. We choose 5 Gauss-Lobatto quadrature nodes per time step, leading to an 8th order IMEX method, which requires only 8 time steps to achieve an accuracy of about 5.8×10^{-5}. We do not coarsen in time, but—impressing everybody with how resilient PFASST is to spatial coarsening—go from 512 degrees-of-freedom on the fine to 32 on the coarse level. We are rewarded with the impressive speedup shown in Fig. 3a: using 8 processors, we are 5.7 times faster than the sequential SDC run. To really drive home the point of how amazing this is, we point out to the reader that this corresponds to 71% parallel efficiency. Even the space-parallel linear solver crowd would have to grudgingly accept such an efficiency as respectable.

However, since the spatial method did not change from the example before, we already know that 32 degrees-of-freedom would have been sufficient to achieve a PDE error of about 10^{-5}. So using 512 degrees-of-freedom on the fine level heavily over-resolves the problem in space. Using only the required 32 degrees-of-freedom on the fine level with a similar coarsening factor of 4 only gives a speedup of 2.7, see Fig. 3b (red curve/diamond markers). While we could probably sneak this into an article, the parallel efficiency of 34% will hardly impress anybody outside of the PinT community.

It is worth noting that better resolution in space on the coarse level does not help (blue curve/square markers). This is because the coarse level does not contribute anything to the convergence of the method anymore. Turning it off completely would even increase the theoretical speedup to about 3.5. Hence, for maximum effect:

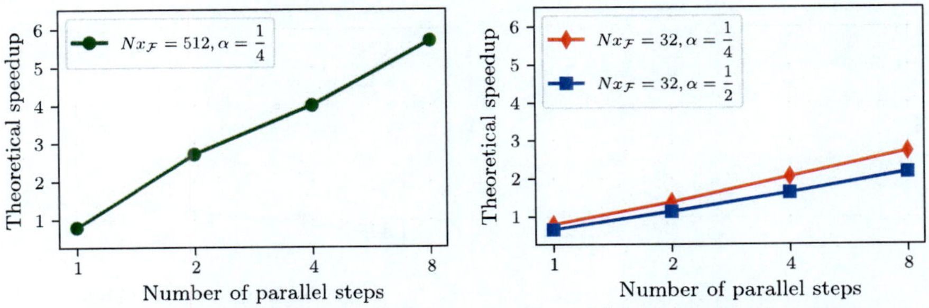

(a) Deceivingly good speedup

(b) Not so good speedup

Fig. 3 Estimated speedup for PFASST runs of the nonlinear Schrödinger example (2) demonstrating the effect of over-resolution in space (*Way* 6)

> **Way 6.** When coarsening in space, make Δx on the fine level so small that even after coarsening, the coarse integrator is accurate. Avoid the temptation to explore a more reasonable resolution.

2.3 Be Smart When Setting Your Iteration Tolerance!

If the audience catches on about your $\Delta t/\Delta x$ over-resolution issues, there is a more subtle way to over-resolve and fool the masses. Since methods like Parareal and PFASST are iterative methods, one must decide when to stop iterating—use this to your advantage! The standard approach is to check the increment between two iterations or some sort of residual (if you can, use the latter: it sounds fancier and people will ask fewer questions). In the runs shown above, Parareal is stopped when the difference between two iterates is below 10^{-10} and PFASST is stopped when the residual of the local collocation problems is below 10^{-10}.

These are good choices, as they give you good speedup: for the PFASST example, a threshold of 10^{-5} would have been sufficient to reach the accuracy of the serial method. While this leads to fewer PFASST iterations (good!), unfortunately, it also makes the serial SDC baseline much faster (bad!). Therefore, with the higher tolerance, speedup looks much less attractive, even in the over-resolved case, see Fig. 4a.

Similarly, when using more reasonable tolerances, the speedup of the well-resolved examples decreases as shown in Fig. 4b. This leads to our next *Way*, which has a long and proud tradition, and for which we can therefore quote Pakin [17] directly,

> **Way 7.** Hence, to demonstrate good [...] performance, always run far more iterations than are typical, necessary, practical, or even meaningful for real-world usage, numerics be damned!

Yet another smart way to over-resolve is to choose a ridiculously small tolerance for an inner iterative solver. Using something cool like GMRES to solve the linear

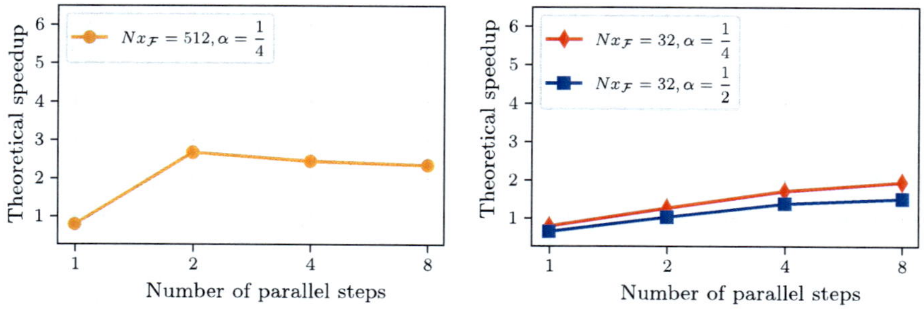

(a) Now also not so good speedup

(b) Still not so good speedup

Fig. 4 Estimated speedup for PFASST runs of the nonlinear Schrödinger example (2) with different resolutions in space demonstrating how using a sensible iteration tolerance of 10^{-5} can reduce speedup (*Way* 7)

systems for an implicit integrator far too accurately is an excellent avenue for making your serial baseline method much slower than it needs to be. This is further desirable because it reduces the impact of tedious overheads like communication or waiting times. We all know that an exceptional way to good parallel performance is using a really slow serial baseline to compare against.

Way 8. Not only use too many outer iterations, but try to maximize the amount of work done by iterative spatial solvers (if you have one, and you always should).

Note that for all the examples presented so far, we did not report any actual speedups measured on parallel computers. Parallel programming is tedious and difficult, as everybody understands, and what do we have performance models for, anyway? It is easier to just plug your parameters into a theoretical model. Realizing this on an actual system can be rightfully considered Somebody Else's Problem (SEP) or a task for your dear future self. But for completeness, the next example will address this directly.

2.4 *Don't Report Runtimes!*

Because solving PDEs only once can bore an audience, we will now talk about optimal control of the heat equation, the "hello world" example in optimization with time-dependent PDEs. This problem has the additional advantage that even more parameters are available to tune. Our specific problem is as follows. Given some desired state u_d on a space-time domain $\Omega \times (0, T)$, $\Omega \subset \mathbb{R}^d$, we aim to find a control c to minimize the objective functional

$$J(u, c) = \frac{1}{2} \int_0^T \|u - u_d\|^2_{L^2(\Omega)} \, \mathrm{d}t + \frac{\lambda}{2} \int_0^T \|c\|^2_{L^2(\Omega)} \, \mathrm{d}t$$

subject to

$$u_t - \nabla^2 u = c + f(u)$$

with periodic boundary conditions allowing us to use FFT to evaluate the Laplacian and perform the implicit linear solves (*Way* 8 be damned). For the linear heat equation considered in the following, the source term is $f(u) \equiv 0$ (see *Way* 1). Optimization is performed using steepest descent; for computation of the required reduced gradient, we need to solve a forward-backward system of equations for state u and adjoint p,

$$\begin{aligned} u_t - \nabla^2 u &= c + f(u) \qquad & -p_t - \nabla^2 p - f'(u)p &= u - u_d \\ u(\cdot, 0) &= 0 & p(\cdot, T) &= 0. \end{aligned}$$

To parallelize in time, we use, for illustration, the most simple approach: given a control c, the state equation is solved parallel-in-time for u, followed by solving the adjoint equation parallel-in-time for p with PFASST using $N_P = 20$ processors. For discretization, we use 20 time steps and three levels with 2/3/5 Lobatto IIIA nodes in time as well as 16/32/64 degrees-of-freedom in space. As a sequential reference, we use MLSDC on the same discretization. We let PFASST/MLSDC iterate until the residual is below 10^{-4} instead of iterating to high precision, so we can openly boast how we avoid *Way* 7.

In the numerical experiments, we perform one iteration of an iterative, gradient-based optimization method to evaluate the method (i.e. solve state, adjoint, evaluate objective, compute gradient). As initial guess for the control we do not use the usual choice $c_0 \equiv 0$ as this would lead to $u \equiv 0$ but a nonzero initial guess—again, we make sure everybody knows that we avoid *Way* 2 in doing so. To estimate speedup, we count PFASST/MLSDC iterations and compute

$$S = \frac{\text{total MLSDC iterations state + adjoint}}{\text{total iterations state on } CPU_{20} + \text{total iterations adjoint on } CPU_1}.$$

We get $S = \frac{40+60}{7+7} = 7.1$, for a nice parallel efficiency of 35.5%.

Before we publish this, we might consider actual timings from a real computer. Unfortunately, using wall clock times instead of iterations gives

$$S = \frac{\text{serial wall clock time}}{\text{parallel wall clock time}} = \frac{44.3\,\text{s}}{18.3\,\text{s}} = 2.4,$$

and thus only roughly a third of the theoretical speedup. To avoid this embarrassment:

Way 9. Only show theoretical/projected speedup numbers (but mention this only in passing or not at all). If you include the cost of communication in the theoretical model, assume it is small enough not to affect your speedup.

Why is the theoretical model poor here? One cause is the overhead for the optimization—after all, there is the evaluation of the objective functional and the construction of gradient. Ignoring parts of your code to report better results is another

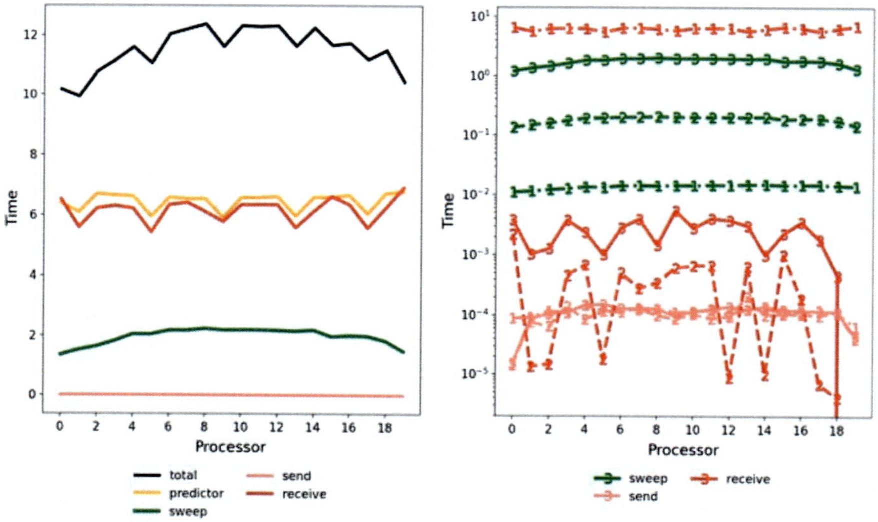

Fig. 5 Wall clock times of the different algorithmic steps for the linear heat equation example on $\Omega = [0, 1]^3$ and $T = 1$. Left: total times. Right: times per level (1 is coarsest level, 3 finest). Note the "receive" times are not negligible as discussed in *Way* 9

proud tradition of parallel computing, see Bailey's paper [3]. However, most of the tasks listed do trivially parallelize in time. The real problem is that communication on an actual HPC system is aggravatingly not really instantaneous.

Looking at detailed timings for PFASST, Fig. 5 shows that the issue truly is in communication costs, which clearly cannot be neglected. In fact, more time is spent on blocking coarse grid communication than on fine sweeps. Note also that, due to the coupled forward-backward solves, each processor requires similar computation and communication times. The following performance model

$$S = \frac{N_P}{\frac{N_P \alpha}{K_S} + \frac{K_P}{K_S}(1 + \alpha + \beta)}$$

accounts for overheads in the β term. Matching the measured speedups requires setting $\beta = 3$ or three times the cost of one sweep on the fine level! This is neither small nor negligible by any measure.

2.5 *Choose the Measure of Speedup to Your Advantage!*

Technically, parallel speedup should be defined as the ratio of the run time of your parallel code to the run time required by the **best** available serial method. But who has the time or energy for such a careful comparison? Instead, it is convenient to choose a baseline to get as much speedup as possible.

In the example above, MLSDC was used as a baseline since it is essentially the sequential version of PFASST and allows for a straightforward comparison and the use of a theoretical speedup. However, MLSDC might not be the fastest serial method to solve state and adjoint equations to some prescribed tolerance. For illustration, we consider solving an optimal control problem for a nonlinear heat equation with $f(u) = -\frac{1}{3}u^3 + u$ on $\Omega \times (0, T) = [0, 20] \times (0, 5)$. Wall clock times were measured for IMEX-Euler, a fourth-order additive Runge-Kutta scheme (ARK-4), and a three-level IMEX-MLSDC with 3/5/9 Lobatto IIIA collocation points, with each method reaching a similar final accuracy in the computed control (thus, using different numbers of time steps, but the same spatial discretization). For the IMEX methods, the nonlinearity as well as the control were treated explicitly. IMEX-Euler was fastest with 102.5 s, clearly beating MLSDC (169.8 s) despite using significantly more time steps. The ARK-4 method here required 183.0 s, as the non-symmetric stage values slow down the forward-backward solves due to the required dense output. With PFASST on 32 CPUs requiring 32 s, the speedup reduces from 5.3 for MLSDC as a reference to 3.2 when compared to IMEX-Euler. By choosing the sequential baseline method wisely, we can increase the reported speedup in this example by more than 65%!

A very similar sleight of hand is hidden in Sect. 2.2, where only theoretical speedups are reported. In the PFASST examples, the SDC iteration counts are used as the baseline results, although in most cases MLSDC required up to 50% fewer iterations to converge. Using MLSDC as a baseline here would reduce the theoretical speedups significantly in all cases. Whether this still holds when actual run times are considered, is, as we have just seen, part of a different story.

Way 10. If you report speedup based on actual timings, compare your code to the method run on one processor and never against a different and possibly more efficient serial method.

2.6 *Use Low-Order Serial Methods!*

A low-order temporal method is a choice convenient for PinT methods because they are easier to implement and allow one to take many time steps without falling prey to *Way* 5, especially when you want to show how the speedup increases as you take ever more time steps for a problem on a fixed time interval. After all, it is the parallel scaling that is exhilarating, not necessarily how quickly one can compute a solution to a given accuracy.

For this example, we will again use Parareal applied to the Kuramoto-Sivashinsky Equation. The K-S equation is a good choice to impress an audience because it gives rise to chaotic temporal dynamics (avoiding *Way* 1). The equation in one dimension reads

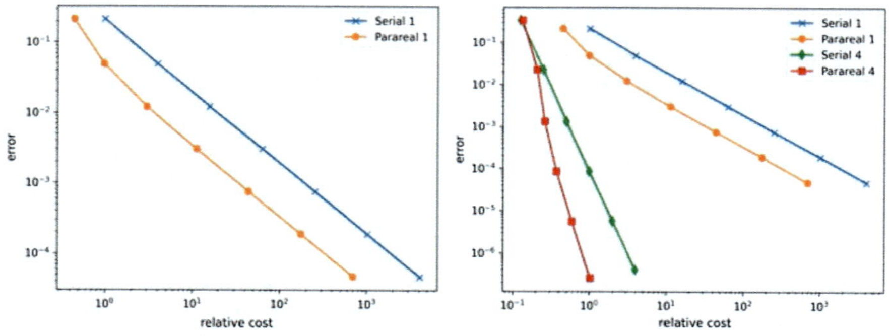

Fig. 6 Comparison of serial and Parareal execution time for the K-S example using a first- and fourth-order ERK integrators. Note that the serial fourth-order integrator is always faster for a given accuracy than the parallel first-order method (*Way* 11)

$$u_t = -uu_x - u_{xx} - u_{xxxx},$$

which we solve on the spatial interval $x \in [0, 32\pi]$ and temporal interval $t \in [0, 40]$. Since the fourth derivative term is linear and stiff, we choose a first-order exponential integrator in a spectral-space discretization where the linear operators diagonalize, and the exponential of the operator is trivial to compute. We use 512 points in space, and this study will compare a serial first-order method with Parareal using the same first-order method in terms of cost per accuracy. Using 32 time processors for all runs, we increase the number of steps for the fine Parareal propagator $\mathcal{F}$ and hence the total number of time steps. The theoretical speedups (ignoring *Way* 9) are displayed on the left panel of Fig. 6. One can see that the Parareal method provides speedup at all temporal resolutions up to a maximum of about 5.85 at the finest resolution (where α is the smallest). So we have achieved meaningful speedup with a respectable efficiency for a problem with complex dynamics. Best to stop the presentation here.

If we are a little more ambitious, we might replace our first-order integrator with the fourth-order exponential Runge-Kutta (ERK) method from [12]. Now we need to be more careful about *Way* 5 and hence won't be able to take nearly as many time steps. In the right panel of Fig. 6, we show the first-order and fourth-order results together. The maximum theoretical speedup attained with the fourth-order method is only about 3.89 at the finest resolution, which is probably reason enough not to do the comparison. But there is the additional irritation that at any accuracy level, the serial fourth-order method is significantly faster than the Parareal first-order method.

Way 11. It is best to show speedup for first-order time integrators since they are a bit easier to inflate. If you want to show speedup for higher-order methods as well, make it impossible to compare cost versus accuracy between first-order and higher-order methods.

3 Parting Thoughts

The careful reader may have noticed that in all the examples above, a single PinT method is used for each *Way*. This brings us finally to

Way 12. Never compare your own PinT method to a different PinT method.

The problem, as we have seen, is that assessing performance for a single PinT method is already not straightforward. Comparing the performance of two or more different methods makes matters even more difficult. Although it has been often discussed within the PinT community, efforts to establish a set of benchmark test examples have, to date, made little headway. The performance of methods like PFASST and Parareal considered here are highly sensitive to the type of equation being solved, the type of spatial discretization being used, the accuracy desired, and the choice of problem and method parameters. In this study, we purposely choose examples that lead to inflated reported speedups, and doing this required us to use our understanding of the methods and the equations chosen. Conversely, in most instances, a simple change in the experiment leads to much worse reported speedups. Different PinT approaches have strengths and weaknesses for different benchmark scenarios, hence establishing a set of benchmarks that the community would find *fair* is a very non-trivial problem.

Roughly, the ways we present can be grouped into three categories: "choose your problem" (*Ways* 1–4), "over-resolve" (*Ways* 5–8), and "choose your performance measure" (*Ways* 9–11). This classification is not perfect as some of the *Ways* overlap. Some of the dubious tricks presented here are intentionally obvious to detect, while others are more subtle. As in the original "twelve ways" article, and those it inspired, the examples are meant to be light-hearted. However, many of the *Ways* have been (unintentionally) used when reporting numerical results, and the authors are not without guilt in this respect. Admitting that, we hope this article will be read the way we intended: as a demonstration of some of the many pitfalls one faces when assessing PinT performance and a reminder that considerable care is required to obtain truly meaningful results.

Acknowledgements The work of Minion was supported by the U.S. Department of Energy, Office of Science, Office of Advanced Scientific Computing Research, Applied Mathematics program under contract number DE-AC02005CH11231. Part of the simulations was performed using resources of the National Energy Research Scientific Computing Center (NERSC), a DOE Office of Science User Facility supported by the Office of Science of the U.S. Department of Energy under Contract No. DE-AC02-05CH11231.

Appendix 1

The value of σ in the first example is 0.02.

References

1. Aktosun, T., Demontis, F., Van Der Mee, C.: Exact solutions to the focusing nonlinear Schrödinger equation. Inverse Problems **23**(5), 2171 (2007)
2. Ascher, U.M., Ruuth, S.J., Spiteri, R.J.: Implicit-explicit Runge-Kutta methods for time-dependent partial differential equations. Appl. Numer. Math. **25**(2), 151–167 (1997)
3. Bailey, D.H.: Twelve ways to fool the masses when giving performance results on parallel computers. Supercomputing Review **4**(8) (1991)
4. Chawner, J.: Revisiting "Twelve ways to fool the masses when describing mesh generation performance". https://blog.pointwise.com/2011/05/23/revisiting-%e2%80%9ctwelve-ways-to-fool-the-masses-when-describing-mesh-generation-performance%e2%80%9d/ (2011). Accessed: 2020-4-28
5. Dongarra, J., et al.: Applied Mathematics Research for Exascale Computing. Tech. Rep. LLNL-TR-651000, Lawrence Livermore National Laboratory (2014). URL http://science.energy.gov/%7E/media/ascr/pdf/research/am/docs/EMWGreport:pdf
6. Emmett, M., Minion, M.L.: Toward an Efficient Parallel in Time Method for Partial Differential Equations. Communications in Applied Mathematics and Computational Science **7**, 105–132 (2012). DOI https://doi.org/10.2140/camcos.2012.7.105
7. Gander, M.J.: 50 years of Time Parallel Time Integration. In: Multiple Shooting and Time Domain Decomposition. Springer (2015). DOI https://doi.org/10.1007/978-3-319-23321-5_3
8. Gear, C.W.: Parallel methods for ordinary differential equations. CALCOLO **25**(1-2), 1–20 (1988). DOI https://doi.org/10.1007/BF02575744
9. Globus, A., Raible, E.: Fourteen ways to say nothing with scientific visualization. Computer **27**(7), 86–88 (1994)
10. Gustafson, J.L.: Twelve ways to fool the masses when giving performance results on traditional vector computers. http://www.johngustafson.net/fun/fool.html (1991). Accessed: 2020-4-28
11. Hoefler, T.: Twelve ways to fool the masses when reporting performance of deep learning workloads! (not to be taken too seriously) (2018). ArXiv, arXiv:1802.09941
12. Krogstad, S.: Generalized integrating factor methods for stiff PDEs. J. Comput. Phys. **203**(1), 72–88 (2005)
13. Lions, J.L., Maday, Y., Turinici, G.: A "parareal" in time discretization of PDE's. Comptes Rendus de l'Académie des Sciences - Series I - Mathematics **332**, 661–668 (2001). DOI https://doi.org/10.1016/S0764-4442(00)01793-6
14. Minhas, F., Asif, A., Ben-Hur, A.: Ten ways to fool the masses with machine learning (2019). ArXiv, arXiv:1901.01686
15. Nievergelt, J.: Parallel methods for integrating ordinary differential equations. Commun. ACM **7**(12), 731–733 (1964). DOI https://doi.org/10.1145/355588.365137
16. Ong, B.W., Schroder, J.B.: Applications of time parallelization. Computing and Visualization in Science **648** (2019)
17. Pakin, S.: Ten ways to fool the masses when giving performance results on GPUs. HPCwire, December **13** (2011)
18. Tautges, T.J., White, D.R., Leland, R.W.: Twelve ways to fool the masses when describing mesh generation performance. IMR/PINRO Joint Rep. Ser. pp. 181–190 (2004)

IMEX Runge-Kutta Parareal for Non-diffusive Equations

Tommaso Buvoli and Michael Minion

Abstract Parareal is a widely studied parallel-in-time method that can achieve meaningful speedup on certain problems. However, it is well known that the method typically performs poorly on non-diffusive equations. This paper analyzes linear stability and convergence for IMEX Runge-Kutta Parareal methods on non-diffusive equations. By combining standard linear stability analysis with a simple convergence analysis, we find that certain Parareal configurations can achieve parallel speedup on non-diffusive equations. These stable configurations possess low iteration counts, large block sizes, and a large number of processors. Numerical examples using the nonlinear Schrödinger equation demonstrate the analytical conclusions.

Keywords Parareal · Parallel-in-time · Implicit-explicit · High-order · Dispersive equations

1 Introduction

The numerical solution of ordinary and partial differential equations (ODEs and PDEs) is one of the fundamental tools for simulating engineering and physical systems whose dynamics are governed by differential equations. Examples of fields where PDEs are used span the sciences from astronomy, biology, and chemistry to zoology, and the literature on methods for ODEs is well established (see e.g. [19, 20]).

Implicit-explicit (IMEX) methods are a specialized class of ODE methods that are appropriate for problems where the right-hand side of the equation is additively

T. Buvoli (✉)
University of California, Merced, Merced, CA 95343, USA
e-mail: tbuvoli@ucmerced.edu

M. Minion
Lawrence Berkeley National Lab, Berkeley, CA 94720, USA
e-mail: mlminion@lbl.gov

© The Author(s), under exclusive license to Springer Nature Switzerland AG 2021
B. Ong et al. (eds.), *Parallel-in-Time Integration Methods*, Springer Proceedings in Mathematics & Statistics 356, https://doi.org/10.1007/978-3-030-75933-9_5

split into two parts so that

$$y'(t) = F^E(y,t) + F^I(y,t). \tag{1}$$

The key characteristic of an IMEX method is that the term F^E (assumed to be non-stiff) is treated explicitly while the term F^I (assumed to be stiff) is treated implicitly. In practice, IMEX methods are often used to solve equations that can be naturally partitioned into a stiff linear component that is treated implicitly and a non-stiff nonlinearity that is treated explicitly. The canonical example is a nonlinear advection-diffusion type equation, where the stiffness comes from the (linear) diffusion terms while the nonlinear terms are not stiff. IMEX methods are hence popular in many fluid dynamics settings. Second-order methods based on a Crank-Nicolson treatment of the diffusive terms and an explicit treatment of the nonlinear terms are a notable example. However, in this study, we consider a different class of problems where the IMEX schemes are applied to a dispersive rather than diffusive term. Here, a canonical example is the nonlinear Schrödinger equation. Within the class of IMEX methods, we restrict the study here to those based on additive or IMEX Runge-Kutta methods (see e.g. [3, 7, 11, 22]). In particular, we will study the behavior of the parallel-in-time method, Parareal, constructed from IMEX Runge-Kutta (hereafter IMEX-RK) methods applied to non-diffusive problems.

Parallel-in-time methods date back at least to the work of Nievergelt in 1964 [27] and have seen a resurgence of interest in the last two decades [17]. The Parareal method introduced in 2001 [24] is perhaps the most well-known parallel-in-time method and can be arguably attributed to catalyzing the recent renewed interest in temporal parallelization. The emergence of Parareal also roughly coincides with the end of the exponential increase in individual processor speeds in massively parallel computers, a development that has resulted in a heightened awareness of the bottleneck to reducing run time for large-scale PDE simulations through spatial parallelization techniques alone. Although Parareal is a relatively simple method to implement (see Sect. 3) and can, in principle, be employed using any single-step serial temporal method, one main theme of this paper is that the choice of method is critical to the performance of the algorithm.

Parareal employs a concurrent iteration over multiple time steps to achieve parallel speedup. One of its main drawbacks is that the parallel efficiency is typically modest and is formally bounded by the inverse of the number of iterations required to converge to the serial solution within a pre-specified tolerance. Another well-known limitation is that the convergence of the method is significantly better for purely diffusive problems than for advective or dispersive ones. As we will show, the convergence properties of the Parareal methods considered here are quite complex, and the efficiency is sensitive to the problem being solved, the desired accuracy, and the choice of parameters that determine the Parareal method. In practice, this makes the parallel performance of Parareal difficult to summarize succinctly.

Incorporating IMEX integrators into Parareal enables the creation of new Parareal configurations that have similar stability and improved efficiency compared to Parareal configurations that use fully implicit solvers. IMEX Parareal integrators were first proposed by Wang et al. [34], where their stability is studied for equations with a stiff dissipative linear operator and a non-stiff, non-diffusive, and nonlinear operator. In this work, we focus exclusively on non-diffusive equations where the spectrums of F^E and F^I are both purely imaginary. Moreover, we only consider Parareal methods on bounded intervals with a fixed number of iterations. Under these restrictions, one can interpret Parareal as a one-step Runge-Kutta method with a large number of parallel stages. By taking this point of view, we can combine classical linear stability and accuracy theory with more recent convergence analyses of Parareal [29]. Furthermore, fixing the parameters means that the parallel cost of Parareal is essentially known a priori making comparisons in terms of accuracy versus wall-clock more straightforward.

The main contribution of this work is to introduce new diagrams that combine convergence regions and classical linear stability regions for Parareal on the partitioned Dahlquist problem. The diagrams and underlying analysis can be used to determine whether a particular combination of integrators and parameters will lead to a stable and efficient Parareal method. They also allow us to identify the key Parareal parameter choices that can provide some speedup for non-diffusive problems. Overall, the results can be quite surprising, including the fact that convergence regions do not always overlap with stability regions; this means that a rapidly convergent Parareal iteration does not imply that Parareal considered as a one-step method with a fixed number of iterations is stable in the classical sense.

The rest of this paper is organized as follows. In the next section, we present a general overview of IMEX-RK methods and the specific methods used in our study are identified. In Sect. 3, we provide a short review of the Parareal method followed by a discussion of the theoretical speedup and efficiency. In Sect. 4, we conduct a detailed examination of the stability and convergence properties of IMEX-RK Parareal methods. Then, in Sect. 5, we present several numerical results using the nonlinear Schrödinger equation to confirm the insights from the linear analysis. Finally, we present a summary of the findings, along with our conclusions in Sect. 6.

2 IMEX Runge-Kutta Methods

In this section, we briefly discuss the IMEX Runge-Kutta (IMEX-RK) methods that are used in this paper. Consider the ODE (1) where F^E, is assumed to be non-stiff and while F^I is assumed to be stiff. Denoting y_n as the approximation to $y(t_n)$ with $\Delta t = t_{n+1} - t_n$, the simplest IMEX-RK method is forward/backward Euler

$$y_{n+1} = y_n + \Delta t \left(F^E(y_n, t_n) + F^I(y_{n+1}, t_{n+1})\right). \tag{2}$$

In each step, one needs to evaluate $F^E(y_n, t_n)$ and then solve the implicit equation

$$y_{n+1} - \Delta t F^I(y_{n+1}, t_{n+1}) = y_n + \Delta t F^E(y_n, t_n). \tag{3}$$

Higher order IMEX methods can be constructed using different families of integrators and IMEX-RK methods (also called *additive* or *partitioned*) are one popular choice (see e.g. [3, 7, 11, 22]). The generic form for an s stage IMEX-RK method is

$$y_{n+1} = y_n + \Delta t \left(\sum_{j=1}^{s} b_j^E F^E(Y_j, t_n + \Delta t c_j^E) + b_j^I F^I(Y_j, t_n + \Delta t c_j^I) \right), \tag{4}$$

where the stage values are

$$Y_j = y_n + \Delta t \left(\sum_{k=1}^{s-1} a_{j,k}^E F^E(Y_k, t_n + \Delta t c_k^E) + \sum_{k=1}^{s} a_{j,k}^I F^I(Y_k, t_n + \Delta t c_k^I) \right). \tag{5}$$

Such methods are typically encoded using two Butcher tableaus that, respectively, contain the coefficients $a_{j,k}^E$, b_j^E, c_j^E and $a_{j,k}^I$, b_j^I, and c_j^I. As with the Euler method, each stage of an IMEX method requires the evaluation of $F^E(y_j, t_j)$, $F^I(y_j, t_j)$ and the solution of the implicit equation

$$Y_j - (\Delta t a_{j,j}^I) F^I(Y_j, t_n + \Delta t c_j^I) = r_j, \tag{6}$$

where r_j is a vector containing all the known quantities that determine the jth stage. IMEX methods are particularly attractive when $F^I(y, t) = Ly$, where L is a linear operator so that (6) becomes

$$(I - \Delta t a_{j,j}^I L) Y_j = r_j. \tag{7}$$

If a fast preconditioner is available for inverting these systems, or if the structure of L is simple, then IMEX methods can provide significant computational savings compared to fully implicit methods.

To achieve a certain order of accuracy, the coefficients a^E and a^I must satisfy both order and matching conditions. Unfortunately, the total number of conditions grows extremely fast with the order of the method, rendering classical order-based constructions difficult. To the best of the authors' knowledge, there are currently no IMEX methods beyond order five that have been derived using classical order conditions. However, by utilizing different approaches, such as extrapolation methods [12] or spectral deferred correction [15, 25], it is possible to construct high-order IMEX methods.

In this work, we consider IMEX-RK methods of order one through four. The first- and second-order methods are the (1, 1, 1) and (2, 3, 2) methods from [3] whose tableaus can be found in Sects. 2.1 and 2.5, respectively. The third- and fourth-order methods are the ARK3(2)4L[2]SA and ARK4(3)6L[2]SA, respectively, from [22]. All the schemes we consider have an L-Stable implicit integrator.

3 The Parareal Method

The Parareal method, first introduced in 2001 [24], is a popular approach to time-parallelization of ODEs. In this section, we will give a brief overview of Parareal and then present a theoretical model for the parallel efficiency and speedup of the method (Table 1).

3.1 Method Definition

In the original form, Parareal is straightforward to describe by a simple iteration. Let $[T_0, T_{fin}]$ be the time interval of interest and t_n denote a series of time steps in this interval. Next, define coarse and fine propagators $\mathcal{G}$ and $\mathcal{F}$, each of which produces an approximation to the ODE at t_{n+1} given an approximation to the solution at t_n.

Assume that one has a provisional guess of the solution at each t_n, denoted y_n^0. This is usually provided by a serial application of the coarse propagator $\mathcal{G}$. Then the kth Parareal iteration is given by

$$y_{n+1}^{k+1} = \mathcal{F}(y_n^k) + \mathcal{G}(y_n^{k+1}) - \mathcal{G}(y_n^k), \tag{8}$$

Table 1 Definitions of variable names used in the description of Parareal

Variable	Meaning	Definition
T_{fin}	Final time of ODE	Problem-specified
N_p	Number of processors	User-defined
N_s	Total Number of fine steps	User-defined
N_b	Number of Parareal blocks	User-defined
K	Number of Parareal iterations	User-defined or adaptively controlled
Δt	Time step for serial method	T_{fin}/N_s
$\mathcal{G}$	Fine propagator	Here 1 step of IMEX RK method
$\mathcal{F}$	Fine propagator	N_f steps of IMEX RK method
N_f	Number of RK steps in $\mathcal{F}$	$N_s/(N_p N_b)$
N_g	Number of RK steps in $\mathcal{G}$	1 for all examples
N_T	Total number of fine steps per block	N_s/N_b
C_s	Cost of full serial run	$N_s c_f$
c_g	Cost of method per step in $\mathcal{G}$	User-defined
c_f	Cost of method per step in $\mathcal{F}$	User-defined
$C_{\mathcal{F}}$	Cost of $\mathcal{F}$	$N_f c_f$
$C_{\mathcal{G}}$	Cost of $\mathcal{G}$	$N_g c_g$
α	Ratio of $\mathcal{G}$ to $\mathcal{F}$ cost	$C_{\mathcal{G}}/C_{\mathcal{F}}$

where the critical observation is that the $\mathcal{F}(y_n^k)$ terms can be computed on each time interval in parallel. The goal of Parareal is to iteratively compute an approximation to the numerical solution that would result from applying $\mathcal{F}$ sequentially on N_p time intervals,

$$y_{n+1} = \mathcal{F}(y_n), \qquad \text{for } \ n = 0 \dots N_p - 1 \tag{9}$$

where each interval is assigned to a different processor. As shown below, assuming that $\mathcal{G}$ is computationally much less expensive than $\mathcal{F}$ and that the method converges in few enough iterations, parallel speedup can be obtained.

A part of the appeal of the Parareal method is that the propagators $\mathcal{G}$ and $\mathcal{F}$ are not constrained by the definition of the method. Hence, Parareal as written can in theory be easily implemented using any numerical ODE method for $\mathcal{G}$ and $\mathcal{F}$. Unfortunately, as discussed below, not all choices lead to efficient or even convergent parallel numerical methods, and the efficiency of the method is sensitive to the choice of parameters.

Note that as described, the entire Parareal method can be considered as a self-starting, single-step method for the interval $[T_0, T_{fin}]$ with time step $\Delta T = T_{fin} - T_0$. In the following section, the classical linear stability of Parareal as a single-step method will be considered for $\mathcal{G}$ and $\mathcal{F}$ based on IMEX-RK integrators. This perspective also highlights the fact that there is a choice that must be made for any particular Parareal runs regarding the choice of ΔT. To give a concrete example for clarity, suppose the user has an application requiring 1024 time steps of some numerical method to compute the desired solution on the time interval [0, 1], and that 8 parallel processors are available. She could then run the Parareal algorithm on 8 processors with 128 steps of the serial method corresponding to $\mathcal{F}$. Alternatively, Parareal could be run as a single step method on two *blocks* of time steps corresponding to [0, 1/2] and [1/2, 1] with each block consisting of 512 serial fine time steps, or 64 serial steps corresponding to $\mathcal{F}$ for each processor on each block. These two blocks would necessarily be computed serially with the solution from the first block at $t = 1/2$ serving as the initial condition on the second block.

3.2 *Cost and Theoretical Parallel Speedup and Efficiency*

We describe a general framework for estimating the potential speedup for the Parareal method in terms of the reduction in the run time of the method. Although theoretical cost estimates have been considered in detail before (see e.g. [4], we repeat the basic derivation for the specific assumptions of the IMEX-RK based methods used here and for the lesser known estimates for multiple block Parareal methods. For simplicity, assume that an initial value ODE is to be solved with some method requiring N_s time steps to complete the simulation on the interval $[0, T_{fin}]$. We assume further that the same method will be used in the fine propagator in Parareal. If each step of the serial method has cost c_f, then the total serial cost is

$$C_s = N_s c_f. \tag{10}$$

In the numerical examples present in Sect. 5, both the coarse and fine propagators consist of a number of steps of an IMEX-RK method applied to a pseudospectral discretization of a PDE. The main cost in general is then the cost of the FFT used to compute explicit nonlinear spatial function evaluations. Hence each step of either IMEX-RK method has essentially a fixed cost, denoted c_f and c_g. This is in contrast to the case where implicit equations are solved with an iterative method and the cost per time step could vary considerably by step.

Given N_p available processors, the Parareal algorithm can be applied to N_b blocks of time intervals, with each block having length $\Delta T = T_{fin}/N_b$. Again for simplicity we assume that in each time block, each processor is assigned a time interval of equal size $\Delta T_p = \Delta T/N_p$. Under these assumptions, $\mathcal{F}$ is now determined to be $N_f = N_s/(N_p N_b)$ steps of the serial method. Parareal is then defined by the choice of $\mathcal{G}$, which we assume here is constant across processors and blocks consisting of N_g steps of either the same or different RK method as used in $\mathcal{F}$ with cost per step c_g. Let $C_{\mathcal{F}} = N_f c_f$ be the time needed to compute $\mathcal{F}$, and likewise, let $C_{\mathcal{G}} = N_g c_g$ be the cost of the coarse propagator.

The cost of K iterations of Parareal performed on a block is the sum of the cost of the predictor on a block, $N_p C_{\mathcal{G}}$, plus the additional cost of each iteration. In an ideal setting where each processor computes a quantity as soon as possible and communication cost is neglected, the latter is simply the $K(C_{\mathcal{F}} + C_{\mathcal{G}})$. Hence, the total cost of Parareal on a block is

$$C_B = N_p C_{\mathcal{G}} + K(C_{\mathcal{F}} + C_{\mathcal{G}}). \tag{11}$$

The total cost of Parareal is the sum over blocks

$$C_p = \sum_{i=1}^{N_b} \left(N_p C_{\mathcal{G}} + K_i(C_{\mathcal{F}} + C_{\mathcal{G}})\right) = N_b N_p C_{\mathcal{G}} + (C_{\mathcal{F}} + C_{\mathcal{G}}) \sum_{i=1}^{N_b} K_i, \tag{12}$$

where K_i is the number of iterations required to converge on block i. Let $\bar{K}$ denote the average number of iterations across the blocks then

$$C_p = N_b \left(N_p C_{\mathcal{G}} + \bar{K}(C_{\mathcal{F}} + C_{\mathcal{G}})\right). \tag{13}$$

Note the first term $N_b N_p C_{\mathcal{G}}$ is exactly the cost of applying the coarse propagator over the entire time interval.

Finally, denoting $\alpha = C_{\mathcal{G}}/C_{\mathcal{F}}$, the speedup $S = C_s/C_p$ is then

$$S = \frac{N_p}{N_p \alpha + \bar{K}(1 + \alpha)}. \tag{14}$$

For fixed K, where $\bar{K} = K$, this reduces to the usual estimate (see e.g. Eq. (28) [26] or Eq. (19) of [4]). The parallel efficiency of Parareal $E = S/N_p$ is

$$E = \frac{1}{(N_p + \bar{K})\alpha + \bar{K}}. \tag{15}$$

A few immediate observations can be made from the formulas for S and E. Clearly, the bound on efficiency is $E < 1/\bar{K}$. Further, if significant speedup is to be achieved, it should be true that $\bar{K}$ is significantly less than N_p and $N_p\alpha$ is small as well. As will be demonstrated later, the total number of Parareal iterations required is certainly problem-dependent and also dependent on the choices of $\mathcal{F}$ and $\mathcal{G}$. It might seem strange at first glance that the number of blocks chosen does not appear explicitly in the above formulas for S and E. Hence, it would seem better to choose more blocks of shorter length so that $\bar{K}$ is minimized. Note however that increasing the number of blocks by a certain factor with the number of processors fixed means that N_f will decrease by the same factor. If the cost of the coarse propagator $C_{\mathcal{G}}$ is independent of the number of blocks (as in the common choice of $\mathcal{G}$ being a single step of a given method, i.e. $N_g = 1$), then α will hence increase by the same factor. Lastly, one can derive the total speedup by also considering the speedup over each block, S_i as

$$S = \frac{1}{\sum_{i=1}^{N_b} \frac{1}{N_b S_i}} = \frac{1}{\frac{1}{N_b}\sum_{i=1}^{N_b} \frac{1}{S_i}}. \tag{16}$$

Finally, we should note that more elaborate parallelization strategies than that discussed above are possible, for example [1, 4, 6, 28].

4 Non-diffusive Dalquist: Stability, Convergence, Accuracy

In this section, we analyze linear stability and convergence properties for IMEX-RK Parareal methods for non-diffusive problems. There have been multiple previous works that have analyzed convergence and stability properties of Parareal. Bal [5] analyzed Parareal methods with fixed parameters, and Gander and Vandewalle [18] studied the convergence of parareal on both bounded and unbounded intervals as the iterations k tends to infinity. More recently, Southworth et al. [31, 32] obtained tight convergence bounds for Parareal applied to linear problems. Specific to stability for non-diffusive equations, Staff and Ronquist [33] conducted an initial numerical study, Gander [16] analyzed the stability of Parareal using characteristics, and Ruprecht [29] studied the cause of instabilities for wave equations. In this work, we will use the work of Ruprecht as a starting point to study stability and convergence for Parareal integrators that are constructed using IMEX-RK integrators.

In this work, we consider the non-diffusive partitioned Dahlquist test problem

$$\begin{cases} y' = i\lambda_1 y + i\lambda_2 y \\ y(0) = 1 \end{cases} \qquad \lambda_1, \lambda_2 \in \mathbb{R}, \tag{17}$$

where the term $i\lambda_1 y$ is treated implicitly and the term $i\lambda_2 y$ is treated explicitly. This equation is a generalization of the Dahlquist test problem that forms the basis of classical linear stability theory [35, IV.2], and the more general equation with $\lambda_1, \lambda_2 \in \mathbb{C}$ has been used to study the stability properties of various specialized integrators [2, 9, 13, 21, 23, 30]. In short, (17) highlights stability for (1) when $F^E(y,t)$ and $F^I(y,t)$ are autonomous, diagonalizable linear operators that share the same eigenvectors, and have a purely imaginary spectrum.

When solving (17), a classical one-step integrator (e.g. an IMEX-RK method) reduces to an iteration of the form

$$y_{n+1} = R(iz_1, iz_2)y_n \quad \text{where} \quad z_1 = h\lambda_1,\ z_2 = h\lambda_2, \tag{18}$$

and $R(\zeta_1, \zeta_2)$ is the stability function of the method. A Parareal algorithm over an entire block can also be interpreted as a one-step method that advances the solution by N_p total time steps of the integrator $\mathcal{F}$. Therefore, when solving (17), it reduces to an iteration of the form

$$y_{(N_p(n+1))} = R(iz_1, iz_2)y_{(N_p n)}. \tag{19}$$

The stability function $R(\zeta_1, \zeta_2)$ plays an important role for both convergence and stability of Parareal, and the approach we take for determining the stability functions and convergence rate is identical to the one presented in [29].

The formulas and analysis presented in the following two subsections pertain to a single Parareal block. Since we will compare Parareal configurations that vary the number of fine steps N_f (so that the fine integrator $\mathcal{F}$ is $\mathcal{F} = f^{N_f}$), it is useful to introduce the *blocksize* $N_T = N_p N_f$ which corresponds to the total number of steps that the integrator f takes over the entire block.

4.1 Linear Stability

The stability region for a one-step IMEX method with stability function $R(\zeta_1, \zeta_2)$ is the region of the complex ζ_1 and ζ_2 plane given by

$$\hat{\mathcal{S}} = \left\{(\zeta_1, \zeta_2) \in \mathbb{C}^2 : |R(\zeta_1, \zeta_2)| \le 1\right\}. \tag{20}$$

Inside $\hat{\mathcal{S}}$ the amplification factor $|R(\zeta_1, \zeta_2)|$ is smaller than or equal to on, which ensures that the time step iteration remains forever bounded. For traditional integrators, one normally expects to take a large number of time steps, so even a mild instability will eventually lead to unusable outputs.

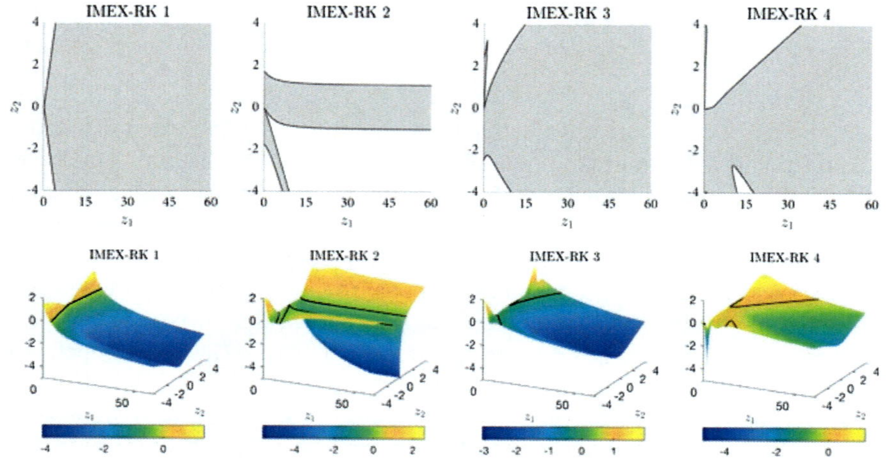

Fig. 1 Non-diffusive linear stability regions (21) for IMEX-RK methods (top) and surface plots showing $\log(\text{amp}(iz_1, iz_2))$ (bottom). For improved readability, we scale the z_1 and z_2 axes differently. For the amplitude function plots, zero marks the cutoff for stability since we are plotting the log of the amplitude function

The full stability region $\hat{\mathcal{S}}$ is four dimensional is difficult to visualize. Since we are only considering the non-diffusive Dahlquist equation, we restrict ourselves to the simpler two dimensional stability region

$$\mathcal{S} = \left\{(z_1, z_2) \in \mathbb{R}^2 : |R(iz_1, iz_2)| \leq 1\right\}. \tag{21}$$

Moreover, all the integrators we consider have stability functions that satisfy

$$R(iz_1, iz_2) = R(-iz_1, -iz_2) \tag{22}$$

which means that we can obtain all the relevant information about stability by only considering $\mathcal{S}$ for $z_1 \geq 0$.

Linear stability for IMEX-RK Before introducing stability for Parareal, we briefly discuss the linear stability properties of the four IMEX-RK methods considered in this work. In Fig. 1, we present 2D stability regions (21) and surface plots that show the corresponding amplitude factor. When $z_2 = 0$, IMEX-RK integrators revert to the fully implicit integrator. Since the methods we consider are all constructed using an L-stable implicit method, the amplification factor will approach zero as $z_1 \rightarrow \infty$. This implies that we should not expect good accuracy for large $|z_1|$ since the exact solution of the non-diffusive Dahlquist equation always has magnitude one. As expected, this damping occurs at a slower rate for the more accurate high-order methods.

Linear stability for Parareal The importance of linear stability for Parareal (i.e. the magnitude of $R(z_1, z_2)$ from (19)) depends on the way the method is run and on the severity of any instabilities. In particular, we consider two approaches for using

Parareal. In the first approach, one fixes the number of processors and integrates in time using multiple Parareal blocks. This turns Parareal into a one-step RK method; therefore, if one expects to integrate over many blocks, then the stability region becomes as important as it is for a traditional integrator.

An alternative approach is to integrate in time using a single large Parareal block. If more accuracy is required, then one simply increases the number of time steps and/or processors, and there is never a repeated Parareal iteration. In this second scenario, we can relax traditional stability requirements since a mild instability in the resulting one-step Parareal method will still produce usable results. However, we still cannot ignore large instabilities that amplify the solution by multiple orders of magnitude.

To analyze the linear stability of parareal, we first require a formula for its stability function. In [29], Ruprecht presents a compact formulation for the stability function of a single Parareal block. He first defines the matrices

$$\mathbf{M}_F = \begin{bmatrix} I & & & \\ -F & I & & \\ & \ddots & \ddots & \\ & & -F & I \end{bmatrix} \qquad \mathbf{M}_G = \begin{bmatrix} I & & & \\ -G & I & & \\ & \ddots & \ddots & \\ & & -G & I \end{bmatrix}, \tag{23}$$

where the constants $F = R_f(iz_1, iz_2)^{N_f}$ and $G = R_c(iz_1, iz_2)^{N_g}$ are the stability functions for the fine propagator $\mathcal{F}$ and the coarse propagator $\mathcal{G}$. The stability function for Parareal is then

$$R(iz_1, iz_2) = \mathbf{c}_2 \left(\sum_{j=0}^{k} \mathbf{E}^j \right) \mathbf{M}_G^{-1} \mathbf{c}_1, \tag{24}$$

where $\mathbf{E} = \mathbf{I} - \mathbf{M}_G^{-1}\mathbf{M}_F^{-1}$ and $\mathbf{c}_1 \in \mathbb{R}^{N_p+1}$, $\mathbf{c}_2 \in \mathbb{R}^{1,N_p+1}$ are $\mathbf{c_1} = [1, 0, \ldots, 0]^T$ and $\mathbf{c_2} = [0, \ldots, 0, 1]$.

4.2 Convergence

A Parareal method will always converge to the fine solution after N_p iterations. However, to obtain parallel speedup, one must achieve convergence in substantially fewer iterations. Convergence rates for a linear problem can be studied by writing the Parareal iteration in matrix form, and computing the maximal singular values of the iteration matrix [29]. Below, we summarize the key formulas behind this observation.

For the linear problem (17), the Parareal iteration (8) reduces to

$$\mathbf{M}_G \mathbf{y}^{k+1} = (\mathbf{M}_G - \mathbf{M}_F)\mathbf{y}^k + \mathbf{b}, \tag{25}$$

where $\mathbf{y} = [y_0, y_1, y_2, \ldots, y_{N_p}]^T$ is a vector containing the approximate Parareal solutions at each fine time step of the integrator $\mathcal{F}$, the matrices $\mathbf{M}_G, \mathbf{M}_F \in \mathbb{R}^{N_p+1,N_p+1}$ are defined in (23) and the vector $\mathbf{b} \in \mathbb{R}^{N_p+1}$ is $[y_0, 0, \ldots, 0]^T$. The Parareal algorithm can now be interpreted as a fixed point iteration that converges to the fine solution

$$\mathbf{y}_F = \left[1, F, F^2, \ldots, F^{N_p}\right]^T y_0 \tag{26}$$

and whose error $\mathbf{e}_k = \mathbf{y}^k - \mathbf{y}_F$ evolves according to

$$\mathbf{e}^k = \mathbf{E}\mathbf{e}^{k-1} \quad \text{where} \quad \mathbf{E} = \mathbf{I} - \mathbf{M}_G^{-1}\mathbf{M}_F. \tag{27}$$

Since Parareal converges after N_p iterations, the matrix $\mathbf{E}$ is nilpotent and convergence rates cannot be understood using the spectrum. However, monotonic convergence is guaranteed if $\|\mathbf{E}\| < 1$ since

$$\|e^{k+1}\| \leq \|\mathbf{E}\|\|e^k\| < \|e^k\|,$$

where $\|\cdot\|$ represents any valid norm. We therefore introduce the *convergence region*

$$\mathcal{C}_p = \left\{(z_1, z_2) : \|E\|_p < 1\right\} \tag{28}$$

that contains the set of all z_1, z_2, where the p-norm of $\mathbf{E}$ is smaller than one and the error iteration (27) is contractive. Note that for rapid convergence that leads to parallel speedup one also needs $\|\mathbf{E}\|_p \ll 1$.

Two-norm for Bounding E In [29], Ruprecht selects $\|\mathbf{E}\|_2 = \max_j \sigma_j$, where σ_j is the jth singular value of $\mathbf{E}$. However, the two-norm needs to be computed numerically, which prevents us from understanding the conditions that guarantee fast convergence.

Infinity-norm for bounding E If we consider the ∞-norm, we can exploit the simple structure of the matrix $\mathbf{E}$ to obtain the exact formula

$$\|\mathbf{E}\|_\infty = \frac{1 - |G|^{N_p}}{1 - |G|}|G - F|. \tag{29}$$

This equality can be obtained directly through simple linear algebra (See Appendix 1) and is similar to the formula used in more sophisticated convergence analysis of Parareal [18, 31] and MGRIT [14]. By using this exact formula, we can understand the requirements that must be placed on the coarse and fine integrators to guarantee a rapidly convergent Parareal iteration. We summarize them in three remarks.

Remark 1 If G is stable, so that its stability function is less then one, and $|G - F| < \frac{1}{N_p}$ then the Parareal iteration converges monotonically. Notice that when $|G| < 1$, then

$$\frac{1-|G|^{N_p}}{1-|G|} = \sum_{j=0}^{N_p-1} |G|^j < N_p. \tag{30}$$

Therefore from (29) it follows that if $|G-F| < \frac{1}{N_p}$ then $\|\mathbf{E}\|_\infty < 1$. Note, however, that this is not always mandatory, and in the subsequent remark we show how this restriction is only relevant for modes with no dissipation. Nevertheless, if we want to add many more processors to a Parareal configuration that converges for all modes, then we also require a coarse integrator that more closely approximates the fine integrator. One way to satisfy this restriction is by keeping N_T fixed while increasing the number of processors; this shrinks the stepsize of the coarse integrator so that it more closely approximates the fine integrator. Another option is to simply select a more accurate coarse integrator or increase the number of coarse steps N_g.

Remark 2 It is more difficult to achieve large convergence regions for a non-diffusive equation than for a diffusive one. If we are solving a heavily diffusive problem $y' = \rho_1 y + \rho_2 y$ where $\mathrm{Re}(\rho_1 + \rho_2) \ll 0$ with an accurate and stable integrator, then $|G| \ll 1$. Conversely, if we are solving a stiff non-diffusive problem (17) with an accurate and stable integrator we expect that $|G| \sim 1$. Therefore,

$$\frac{1-|G|^{N_p}}{1-|G|} \sim \begin{cases} 1 & \text{Diffusive Problem,} \\ N_p & \text{Non-Diffusive Problem.} \end{cases} \tag{31}$$

From this, we see that the non-diffusive case is inherently more difficult since we require that the difference between the coarse integrator and the fine integrator should be much smaller than $\frac{1}{N_p}$ for fast convergence. Moreover, any attempts to pair an inaccurate but highly stable coarse solver ($|G| \ll 1$) with an accurate fine solver ($|F| \sim 1$) will at best lead to slow convergence for a non-diffusive problem since $|G-F| \sim 1$. Rapid convergence is possible if both $|F| \ll 1$ and $|G| \ll 1$, however, this is not meaningful convergence since both the coarse and fine integrator are solving the non-diffusive problem inaccurately.

Remark 3 If G is not stable (i.e. $|G| > 1$), then fast convergence is only possible if F is also unstable so that $|F| > 1$. Convergence requires that the difference between the coarse and fine iterator is sufficiently small so that

$$|G-F| < \frac{1-|G|}{1-|G|^{N_p}}. \tag{32}$$

Since G and F are complex numbers we can interpret $\frac{1-|G|}{1-|G|^{N_p}}$ as the radial distance between the numbers. If we want $|F| \leq 1$, then G can never be more than the distance $\frac{1-|G|}{1-|G|^{N_p}}$ from the unit circle. Therefore, we require that

$$|G| - \frac{1 - |G|}{1 - |G|^{N_p}} \leq 1 \quad \Longrightarrow \quad |G| \leq 1. \tag{33}$$

4.3 Linear Stability and Convergence Plots for Parareal

The aim of this subsection is to broadly categorize the effect that each of the Parareal parameters has on stability and convergence. The first parameters that we consider are the coarse and fine integrators. Since we are considering IMEX-RK methods with orders one to four, there are ten possible IMEX-RK pairings where the fine integrator has higher or equivalent order compared to the coarse integrator. The remaining parameters are the number of processors, the number of Parareal iterations K, and the number of coarse and fine integrator steps, N_g and N_f.

In practice, Parareal is commonly run with an adaptively selected K that causes the method to iterate until a pre-specified residual tolerance is satisfied. When discussing linear stability, we instead assume that Parareal always performs a fixed number of iterations. This simplifies our analysis and allows us to quantify how iteration count affects stability. Since the matrix $\mathbf{E}$ from (27) does not depend on K, the number of iterations has no effect on the convergence regions. Therefore, our convergence results apply to both non-adaptive and adaptive Parareal.

To reduce the number of total parameters further, we always take the number of coarse integrator steps N_g to be one. Even with this simplification, there are still too many degrees-of-freedom to discuss all the resulting Parareal methods. Therefore, we only show several example plots that capture the essential phenomena and provide a set of general remarks to encapsulate our main observations. Those who wish to see additional stability and convergence plots can download our Matlab code [10] which can be used to generate figures for any other set of Parareal parameters.

To showcase the stability and convergence properties of Parareal, we present 2D plots that overlay the linear stability region (21) and the convergence region (28) with $p = \infty$. We start with a simple example that demonstrates how we construct our plots. Consider the Parareal integrator with IMEX-RK3 and IMEX-RK4 as the coarse and fine integrator, respectively, and with parameters $K = 2, N_g = 1, N_f = 8, N_p = 64$. The following three figures show the convergence region (left), the stability region (middle), and an overlay of both (right). Each plot shows the (z_1, z_2) plane where we only consider $z_1 \geq 0$ due to the symmetry condition (22).

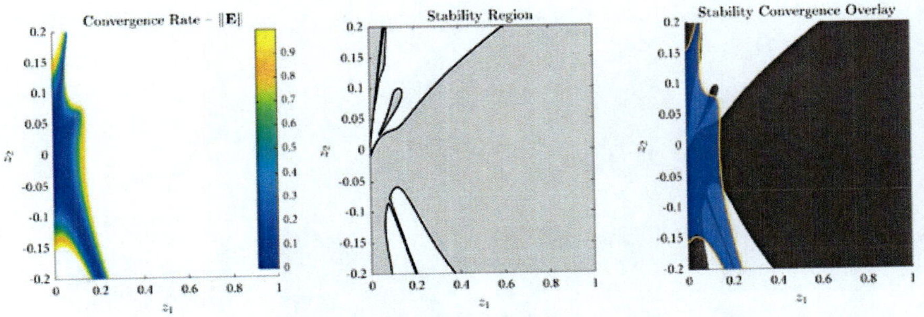

Color	Set	Description
Light Blue	$\mathcal{C}_\infty \setminus \mathcal{S}$	Contractive iteration, unstable one-step method.
Dark Blue	$\mathcal{S} \cap \mathcal{C}_\infty$	Contractive iteration, stable one-step method.
Dark Gray	$\mathcal{S} \setminus \mathcal{C}_\infty$	Non-contractive iteration, stable one-step method.
Yellow	$\|\mathbf{E}\|_\infty = 1$	Boundary of convergence region.

The convergence rate plot has a color bar that corresponds to the norm of the Parareal iteration matrix **E** from (27). In the overlay plot, there are three distinct regions that are colored according to the legend shown above.

To simplify the comparison between different Parareal methods, we scale all the stability functions used to compute both stability and convergence relative to the number of total fine integrator steps N_T; in other words, our plots are generated using the scaled stability function $\hat{R}(iz_1, iz_2) = R(iN_T z_1, iN_T z_2)$.

For brevity, we only present plots for Parareal methods with IMEX-RK3 and IMEX-RK4 as the coarse and fine integrators, and with the following parameters:

$$N_T \in \{512, 2048\}, \quad N_g \in \{1\}, \quad N_f \in \{4, 8, 16, 32\}, \quad k \in \{1, 2, 3, 4\}.$$

For each configuration, the number of processors $N_p = N_T/N_f$. In Figs. 2 and 3, we show a grid of convergence plots and stability-convergence overlay plots for IMEX Parareal integrators with $N_T = 512$ and $N_T = 2048$. Due to the limited stability of the Parareal methods near the origin, we magnify the axes in comparison to our plots for IMEX-RK methods. Three additional figures for Parareal methods with different coarse and fine integrators are shown in Appendix 2, and additional plots with further methods and different axes are available in [10].

General remarks on stability and convergence We make several remarks based on the stability plots shown in Figs. 2, 3, and the additional ones contained in [10]. These remarks are meant to be general and should hold for parareal methods constructed using a larger range of IMEX-RK families than the four discussed in this work.

1. There are many regions in the (z_1, z_2) plane where stability and convergence regions do not overlap (in Figs. 2 and 3 these regions appear in a light blue color). Therefore, a Parareal method that does not take $K = N_p$ can be unstable even for (z_1, z_2) pairs that are inside the convergence region. If this Parareal method is iterated over many blocks, or if the instability is large, then the final output will

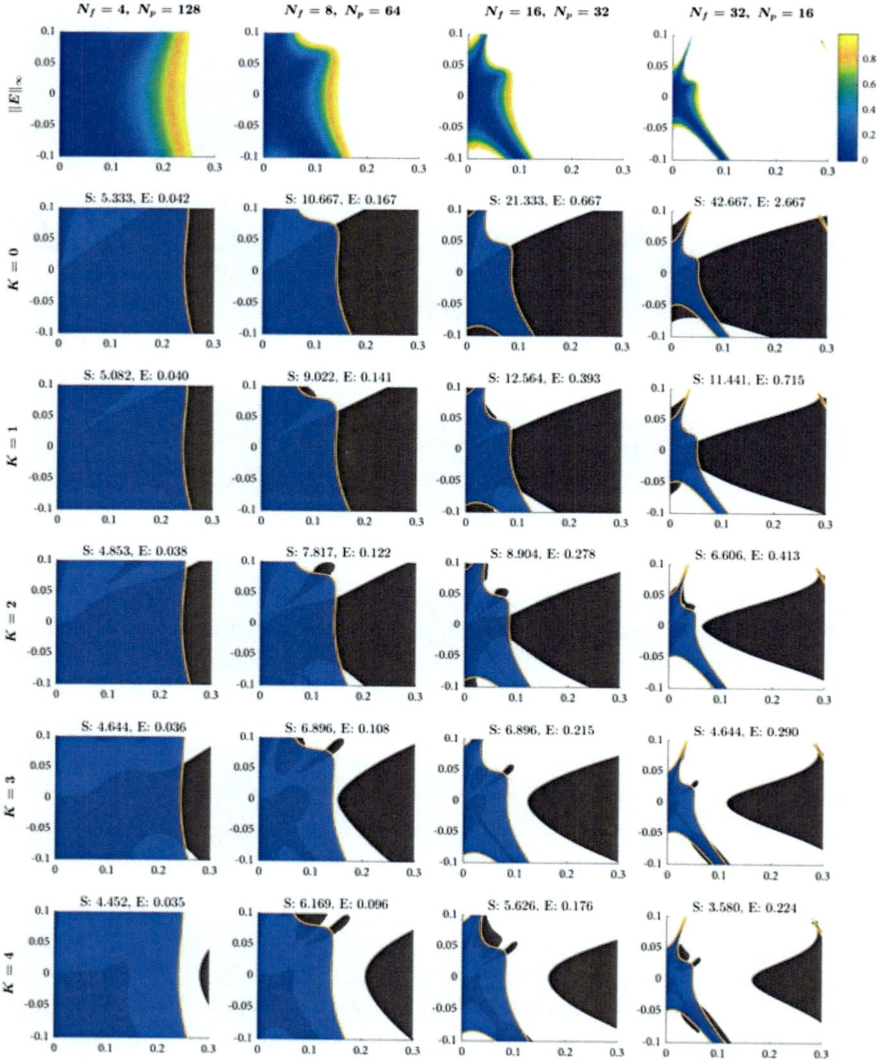

Fig. 2 Stability and convergence overlay plots for Parareal configurations with a block size of $N_T = 512$ and IMEX-RK3, IMEX-RK4 as the coarse and fine integrator. The number of processors increases as one moves rightward in the horizontal direction. The top row shows the convergence rate plot for the method, and all subsequent rows show stability convergence overlays for an increasing number of iterations. The subfigure titles show the theoretical speedup (S) and efficiency (E) for each Parareal configuration

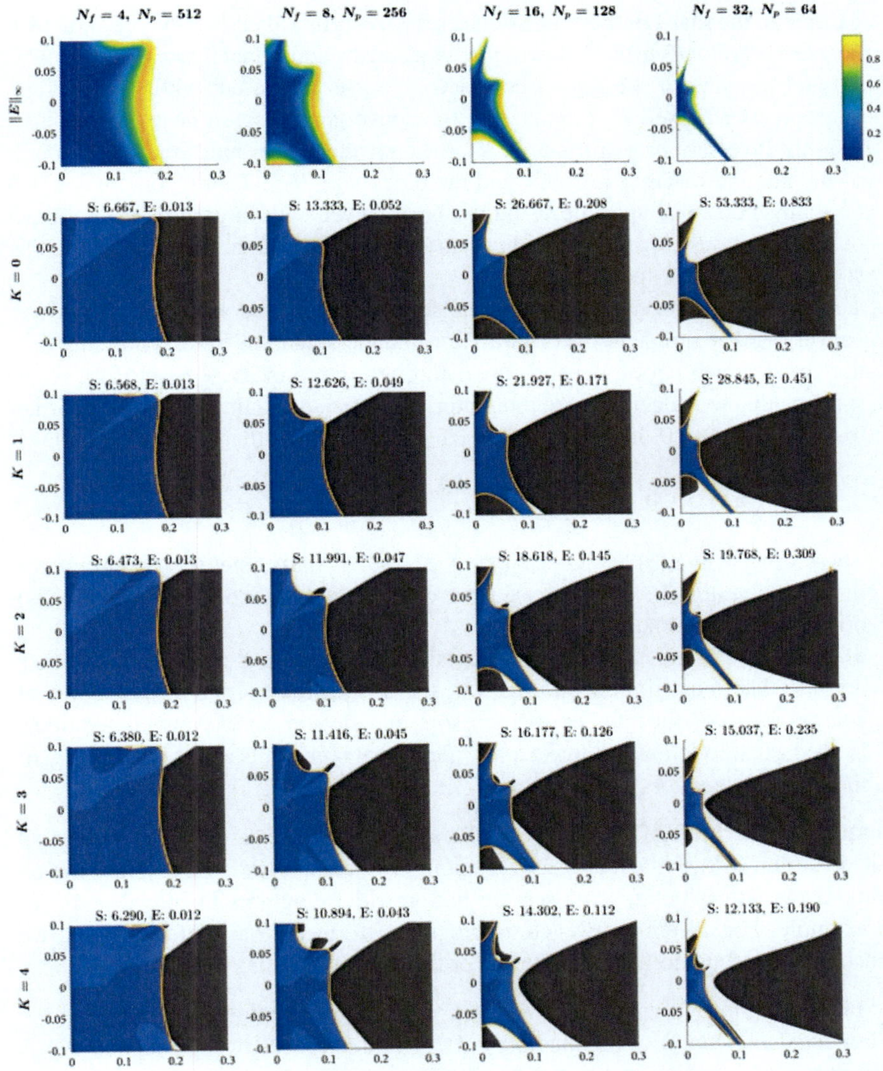

Fig. 3 Stability and convergence overlay plots for Parareal configurations with a block size of $N_T = 2048$ and IMEX-RK3, IMEX-RK4 as the coarse and fine integrator

be unusable. Conversely, lack of convergence paired with stability means that the solution will remain bounded. However, the accuracy for these (z_1, z_2) pairs will never surpass that of the coarse integrator unless the number of iterations is large; in many cases as large as $K = N_p$.

2. Convergence regions for Parareal change drastically depending on the parameters. Overall, we see that the choice of coarse integrator and the number of processors have the most prominent effect on the convergence regions near the origin. In

particular, the shape of the convergence region is primarily driven by the choice of coarse integrator (See the following remark and the additional figures in Appendix 2), and the size of the region contracts if N_T is increased, while N_f is kept constant. For a fixed N_T, the size of the convergence region near the origin is roughly inversely proportional to N_f (or equivalently, proportional to the ratio of the number of processors N_p and the block size N_T). Though one may hope to attain the fastest speedup using a coarse solver with large time steps, these observations suggest that it is important to carefully balance speedup with the size of the solution spectrum.

3. The choice of fine integrator has almost no effect on the shape or size of the convergence region near the origin. We can understand this by first noticing that the convergence region depends on F through the term $|G - F|$ in (29). Now, suppose that we select a coarse integrator of order γ, and a fine integrator of order $\gamma + \delta$, where $\delta \geq 0$. If we define $\lambda = z_1 + z_2$, it follows that

$$G = \exp(i\lambda) + \mathcal{O}\left(|\lambda|^{\gamma}\right), \qquad F(\delta) = \exp(i\lambda) + \mathcal{O}\left(|\lambda|^{\gamma+\delta}\right).$$

Therefore, $|G - F(\delta)| = |G - F(0)| + \mathcal{O}\left(|\lambda|^{\gamma}\right)$. This implies that for any choice of fine integrator, the convergence rates will be nearly identical near the (z_1, z_2) origin where λ is small.

4. Stability regions are non-trivial and depend on more parameters than convergence regions. The most significant effect on stability regions is due to the choice of coarse integrator, the number of processors, and the number of Parareal iterations. In the next subsection, we make a few additional comments about the stability of the IMEX integrators tested in this work.

Remarks regarding IMEX Parareal Overall, IMEX Parareal methods are not well suited for non-diffusive equations. Our main observation is that the stability region splits in two along the z_1 axis; see the bottom right diagrams in Figs. 2 and 3 as an example. For all ten IMEX-RK pairings that form the basis of this work, we consistently find the following patterns for the stability regions near the origin:

- Stability regions grow larger if N_T is increased while keeping N_f constant.
- For fixed N_T, stability regions grow if N_f is decreased (or equivalently N_p is increased).
- Stability regions do not monotonically increase with iteration count. Instead, the regions initially contract and separate as K increases.

Nevertheless, IMEX Parareal methods can still be effective for solving non-diffusive equations under three conditions:

1. **Select a stable method pairing.** This is non-trivial and many pairings lead to an IMEX Parareal method with no stability along the z_1 axis for small K. Of the ten possible coarse-fine integrator pairings, only the ones with IMEX-RK1 or IMEX-RK3 as the coarse integrator were stable. Selecting IMEX-RK2 or IMEX-RK4 as the coarse integrator produced an unstable Parareal method for all the parameters

we considered. More generally, this suggests that RK pairs should be specially constructed to maximize stability.

2. **Keep the iteration counts low, choose large block lengths, and avoid using too few processors.** Increasing the number of Parareal iterations causes the stability region to separate along the z_1 axis. After looking across a wide range of parameters, we find that the stability regions of the IMEX integrators consistently get worse as the number of iterations initially increases. This is especially true if we consider small block lengths N_T or small N_p. This suggests that adaptive implementations of IMEX Parareal on non-diffusive problems will likely lead to an unstable method if the residual tolerance is set too low. However, if one fixes the maximum number of iteration K, selects a large block length N_T, and uses a sufficient number of processors N_p, then the methods can be effective.
   We note that choosing large block lengths with many processors will limit the maximal theoretical speedup of the parareal method. However, for non-diffusive equations, optimizing purely for parallel speedup will lead either to an unstable method or to a parareal method with a very small convergence region that renders it less efficient compared to the serial fine integrator. Moreover, the important metric is not parallel speedup, but rather efficiency (error vs time) of the parareal integrator compared to its serial fine integrator. To properly determine whether a parareal configuration is efficient, one needs to carefully balance stability, speedup, and convergence relative to the spectrum of the ODE problem that is being solved.
3. **Avoid problems with broad spectrums**. Stable pairings of IMEX integrators possess good stability and convergence near the (z_1, z_2) origin. However, for all the RK pairings we tested, the convergence regions do not extend far along the z_1 axis. For moderately sized z_1, we consistently see a region of good stability that is paired with non-contractive convergence. These observations suggest that IMEX Parareal methods will converge slowly on solutions where the energy is concentrated in these modes. We note that the convergence region is restored as z_1 gets sufficiently large (outside the range of the figures contained in the paper). However, in these regions the coarse and fine IMEX integrators both exhibit heavy damping, therefore, rapid convergence to the fine integrator is not a sign of good accuracy; see Remark 2 in Sect. 4.2.

4.4 Accuracy Regions for the Non-diffusive Dahlquist Equation

To supplement our stability plots, we also consider the accuracy regions for IMEX integrators. The accuracy region $\mathcal{A}_\epsilon$ shows the regions in the (z_1, z_2) plane where the difference between the exact solution and the numerical method is smaller than ϵ. The accuracy region for an IMEX method is typically defined as

$$\mathcal{A}_\epsilon = \{(z_1, z_2) \in \mathbb{R} : |R(iz_1, iz_2) - \exp(iz_1, iz_2)| \leq \epsilon\}, \tag{34}$$

where $R(\zeta_2, \zeta_2)$ is the stability function of the method. Since Parareal methods advance the solution multiple time steps, we scale the accuracy regions by the total number of time steps in a block so that

$$\mathcal{A}_\epsilon = \{(z_1, z_2) \in \mathbb{R} : |R(iN_T z_1, iN_T z_2) - \exp(iN_T z_1, iN_T z_2)| \leq \epsilon\}. \quad (35)$$

Under this scaling, the accuracy region of a fully converged Parareal method with $K = N_p$ is

$$\mathcal{A}_\epsilon = \left\{(z_1, z_2) \in \mathbb{R} : |R_F(iz_1, iz_2)^{N_T} - \exp(iN_T z_1, iN_T z_2)| \leq \epsilon\right\}. \quad (36)$$

where $R_F(\zeta_1, \zeta_2)$ is the stability function of the fine integrator; this is the accuracy region of a method that consists of N_T steps of the fine integrator.

In Fig. 4, we show the accuracy regions for IMEX Parareal methods with $N_T = 2048$. The diagrams highlight the importance of the different regions in our stability plots. First, we can clearly see fast convergence to the fine solution inside contractive regions where $\|\mathbf{E}\| \ll 1$. Second, for any (z_1, z_2) pair that are inside the stability region but outside the convergence region, we see that increasing the number of Parareal iterations does not improve the accuracy of the solution beyond that of the coarse integrator. Finally, in the regions with large instabilities, the solution is no longer useful. However, many of the instabilities that lie inside the convergence regions are so small in magnitude that they do not appear prominently in the accuracy plots since they will only affect accuracy after many steps of Parareal.

5 Numerical Experiments

The aim of our numerical experiments is to validate the results from linear stability analysis and demonstrate that we can obtain meaningful parallel speedup with IMEX-RK Parareal on a nonlinearly dispersive equation. In particular, we show that

1. Taking too many Parareal iterations leads to an unstable method; however, selecting larger block sizes N_T for a fixed N_f increases the range of stable choices for the iteration count K.
2. Decreasing the number of processors for a fixed block size N_T and fixed N_g eventually leads to an unstable method, even for time steps where the coarse integrator is stable.
3. For the parameters considered in this work, Parareal methods with an adaptive iteration count K are effective so long as one limits the maximum number of iterations.

For our model nonlinear problem, we select the one dimensional nonlinear Schrödinger (NLS) equation

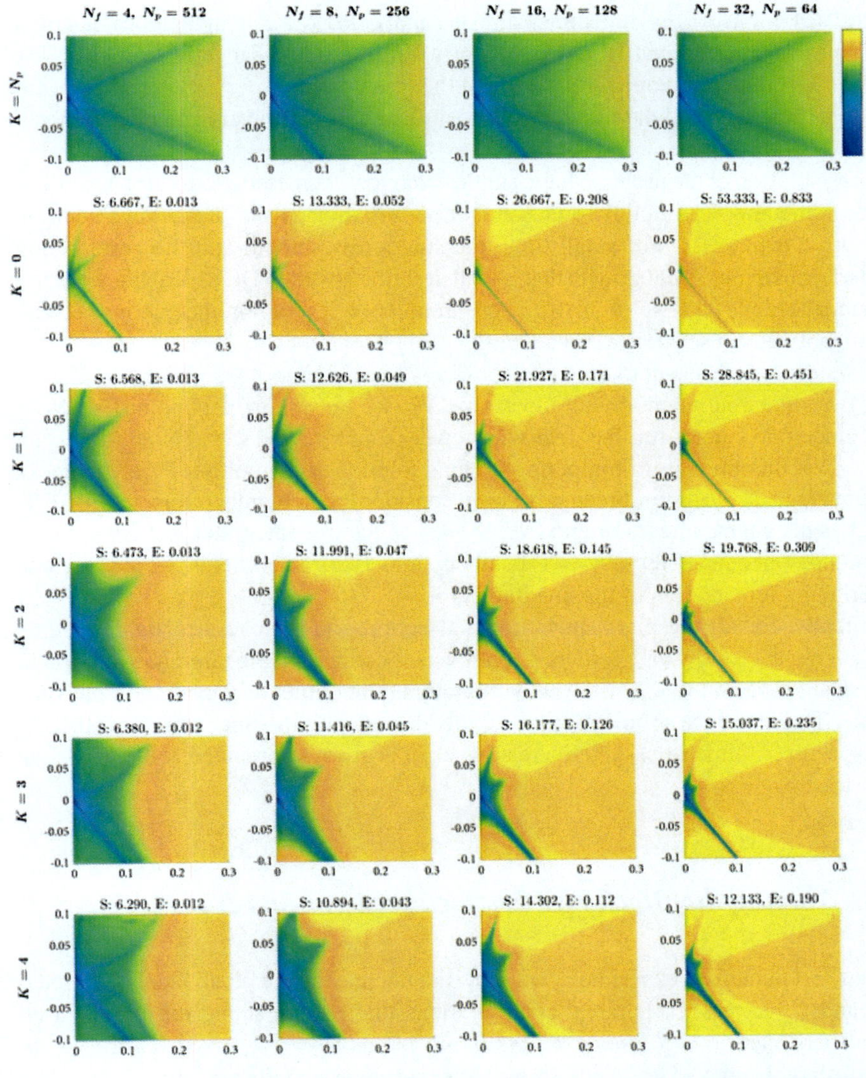

Fig. 4 Accuracy regions for Parareal with a block size of $N_T = 2048$ and IMEX-RK3, IMEX-RK4 as the coarse and fine integrators. These plots complement the stability plots in Fig. 3. The top row shows the accuracy of the fine integrator, the second row shows the coarse integrator, and all subsequent rows show parareal methods with an increasing number of iterations. Color represents the log of the absolute value of the error

$$\begin{aligned} &iu_t + u_{xx} + 2|u|^2 u = 0, \\ &u(x, t = 0) = 1 + \tfrac{1}{100}\exp(ix/4), \quad x \in [-4\pi, 4\pi]. \end{aligned} \tag{37}$$

We equip NLS with periodic boundary conditions and use the method of lines with a fixed spatial grid. For the spatial component, we use a Fourier spectral discretization with 1024 points. In the time dimension, we apply IMEX Parareal methods that treat the linear spatial derivative term implicitly and the nonlinear term explicitly. We solve the equation in Fourier space where the discrete linear derivative operators are diagonal matrices and the implicit solves amount to multiplications with a diagonal matrix.

The solution of the problem has a subtle behavior when used as a test for a parallel-in-time method. The initial condition is a perturbation of the plane wave solution $u(t, x) = \exp(-2it)$. For small time, the solution is smooth and the perturbation experiences exponential growth that is well described by the PDE's local linearization around the plane wave solution [8]. Then, around $t = 10$, the nonlinear terms become dominant and cause spectral broadening.

We integrate (37) out to time $t = 15$ using $N_s = 2^p$ time steps where $p = 7, \ldots, 18$. For a Parareal method with block size N_T, we only show data points for $N_s \geq N_T$ since one cannot run Parareal with fewer than N_T time steps. Note that when $N_s > N_T$, this amounts to computing the final solution using multiple Parareal blocks where $N_b = N_s/N_T$. For brevity, we only consider Parareal integrators with IMEX-RK3 as the coarse integrator and IMEX-RK4 as the fine integrator and always take the number of coarse steps $N_g = 1$. In all our numerical experiments, we also include a serial implementation of the fine integrator.

Finally, for all the plots shown in this section, the relative error is defined as $\|\mathbf{y}_{\text{ref}} - \mathbf{y}_{\text{method}}\|_\infty / \|\mathbf{y}_{\text{ref}}\|_\infty$, where $\mathbf{y}_{\text{ref}}$ is a vector containing the reference solution in physical space and $\mathbf{y}_{\text{method}}$ is a vector containing the output of a method in physical space. The reference solution was computed by running the fine integrator (IMEX-RK4) with 2^{19} time steps, and the relative error is always computed at the final time $t = 15$.

5.1 Varying the Block Size N_T for Fixed N_f and N_g

In our first numerical experiment, we show that increasing the block size N_T for fixed N_f and N_g allows for Parareal configurations that remain stable for an increased number of iterations K. Since we are fixing N_f and N_g, we are increasing N_p to obtain larger block sizes. Therefore, this experiment simultaneously validates the improvement in stability seen when comparing the ith column of Fig. 3 with the ith column of Fig. 2, along with the decrease in stability seen when moving downward along any column of Fig. 2 or 3.

In Fig. 5, we present plots of relative error versus stepsize for three Parareal configurations with block sizes $N_T = 512$, 1024, or 2048. Each of the configurations takes a fixed number of Parareal iterations K and has $N_g = 1$, $N_f = 16$. The stability regions for the Parareal configurations with $N_T = 512$ and $N_T = 2048$ are shown in the third columns of Figs. 2 and 3. From linear stability, we expect that the two Parareal methods will, respectively, become unstable if $K > 3$ and $K > 4$. The experiments with $N_T = 512$ align perfectly with the linear stability regions. For the larger block size, the instabilities are milder and we would need to take many more

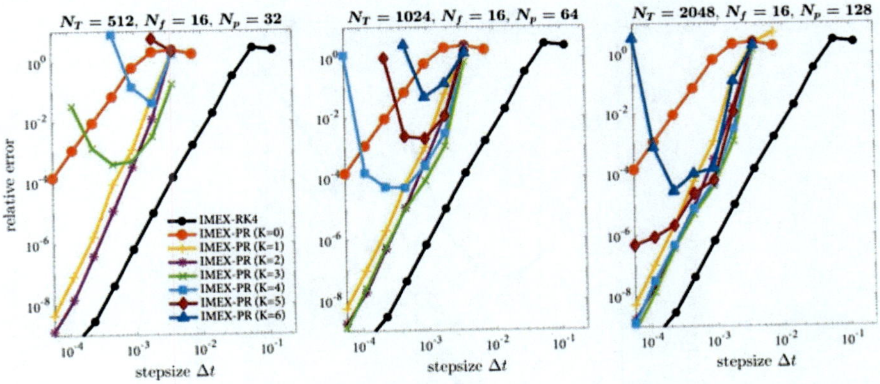

Fig. 5 Accuracy versus stepsize plots for the Parareal method with IMEX-RK3, IMEX-RK4 as the coarse and fine integrator. The block size N_T for the left, middle, and right plots is, respectively, 512, 1024, and 2048. The black line shows the serial fine integrator, while the colored lines represent Parareal methods with different values of iteration count K. Note that the Parareal configuration with $N_T = 512$ and $K = 6$ did not converge for any of the time steps

Parareal blocks for them to fully manifest. Nevertheless, the methods fail to become more accurate for $K = 4$ and start to diverge for $K > 4$. Overall, the results confirm that increasing the block size by increasing N_p leads to an improvement in stability that allows for a larger number of total Parareal iterations.

Finally, we note that an alternative strategy for increasing N_T is to increase N_f while keeping N_p constant. However, we do not consider this scenario since increasing N_f will lead to a method with a significantly smaller convergence region and no better stability (e.g. compare column 2 of Fig. 2 with column 4 of Fig. 3).

5.2 *Varying the Number of Processors N_p for a Fixed N_T and N_g*

In our second numerical experiment, we show that decreasing the number of processors N_p (or equivalently increasing the number of fine steps N_f) for a fixed N_T and N_g will lead to an unstable Parareal method. This experiment validates the stability changes that occur along any row of Fig. 2 or 3.

For brevity, we only consider four Parareal methods with $N_g = 1$, $N_T = 512$, $K = 3$, and $N_p = 16$, 32, 64, or 128. The linear stability regions for each of these methods are shown in the fourth row of Fig. 2. Only the Parareal method with $N_p = 128$ is stable along the entire z_2 axis. The method with $N_p = 64$ has a mild instability located inside its convergence region, and methods with $N_p = 32$ or $N_p = 16$ have large instabilities.

In Fig. 6, we show an accuracy versus stepsize plot that compares each of the four Parareal configurations (shown in colored lines) to the serial fine integrator

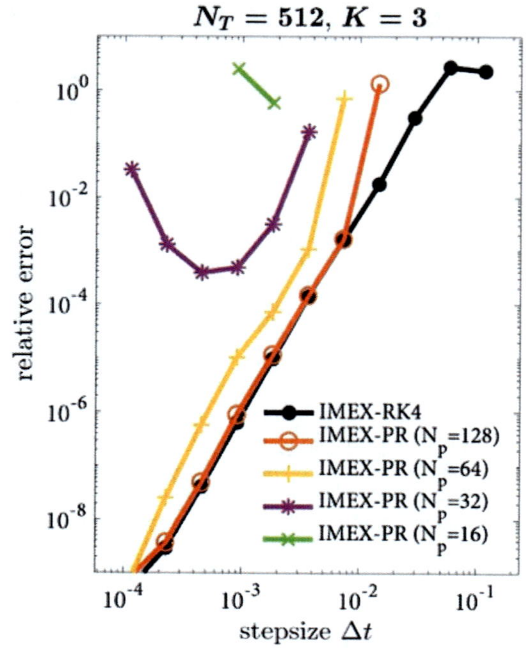

Fig. 6 Variable N_p results

(shown in black). The total number of steps N_s is given by 2^p, where for Parareal $p = 9, \ldots, 18$, while for the serial integrator $p = 7, \ldots, 18$. We can clearly see that decreasing N_p leads to instability even at small stepsizes. Note that for $N_p = 64$ the instability is so small that it does not affect convergence in any meaningful way.

It is important to remark that the largest stable stepsize for any Parareal method is restricted by the stability of its coarse integrator. Since we are taking $N_g = 1$, the number of coarse time steps per block is N_p. Therefore, a Parareal configuration with a smaller N_p takes larger coarse stepsizes and requires a smaller Δt to remain stable. This effect can be seen in Fig. 6, since it causes the rightmost point of the parareal convergence curves to be located at smaller stepsizes Δt for methods with smaller N_p. What is more interesting, however, is that the Parareal configurations with fewer processors are unstable, even when the stepsize is lowered to compensate for the accuracy of the coarse solver. In other words, instabilities form when the difference in accuracy between the coarse and fine solver is too large, even if the coarse solver is sufficiently stable and accurate on its own.

5.3 *Efficiency and Adaptive K*

In our final numerical experiment, we first compare the theoretical efficiency of several IMEX Parareal configurations and then conduct a parallel experiment using the most efficient parameters. We conduct our theoretical efficiency analysis using Parareal methods with $N_T = 2048$, $N_f = 16$, $N_g = 1$, and $K = 1, \ldots, 6$. As shown in the third column of Fig. 4, these configurations possess good speedup and stability regions when $K \leq 3$. To determine the *theoretical runtime* for Parareal (i.e. the runtime in the absence of any communication overhead), we divide the runtime of the fine integrator by the Parareal speedup that is computed using (14).

In Fig. 7a, we show plots of relative error versus theoretical runtime for the seven parareal configurations. The total number of steps N_s is given by 2^p where for Parareal $p = 12, \ldots, 18$, while for the serial integrator $p = 7, \ldots, 18$. Note that the running times for the fine integrator and the parareal configuration with $K = 0$ (i.e. the serial coarse method) measure real-world efficiency since there are no parallelization possibilities for these methods.

The efficiency plots demonstrate that it is theoretically possible to achieve meaningful parallel speedup using IMEX Parareal on the nonlinear Schrödinger equation. Moreover, amongst the seven Parareal configurations, the one with $K = 3$ is the most efficient over the largest range of time steps. However, communication costs on real hardware are never negligible, and real-world efficiency will depend heavily on the underlying hardware and the ODE problem. To validate the practical effectiveness of IMEX Parareal, we ran the most efficient configuration with $K = 3$ in parallel on a distributed memory system with 128 processors.[1] We also tested in identical Parareal configuration with an adaptive controller for K that iterates until either $K \geq K_{max}$ or a residual tolerance of 1×10^{-9} is satisfied. Unsurprisingly, it was necessary to restrict the maximum number of adaptive Parareal iterations to $K_{max} = 3$ or the adaptive controller caused the method to become unstable.

In Fig. 7b, we show plots of relative error versus parallel runtime, and in Table 2, we also include the corresponding speedup for the two Parareal methods. Even on this simple 1D problem, we were able to achieve approximately a ten-fold real-world speedup relative to the serial IMEX-RK4 integrator. This is very encouraging since the ratio between the communication and time step costs is larger for a 1D problem. Our results also show that there is not much noticeable difference between the parareal method with fixed K and the method with adaptive K, except at the finest time steps where the adaptive implementation is able to take fewer iterations.

6 Summary and Conclusions

We have introduced a methodology for categorizing the convergence and stability properties of a Parareal method with pre-specified parameters. By recasting the Parareal algorithm as a one-step RK method with many stages, we are able to combine classical stability analysis with a simple bound on the norm of the Parareal iteration matrix. The resulting stability convergence overlay plots highlight the key

[1] The numerical experiments were performed on the Cray XC40 "Cori" at the National Energy Research Scientific Computing Center using four 32-core Intel "Haswell" processor nodes. The Parareal method is implemented as part of the open-source package LibPFASST available at https://github.com/libpfasst/LibPFASST.

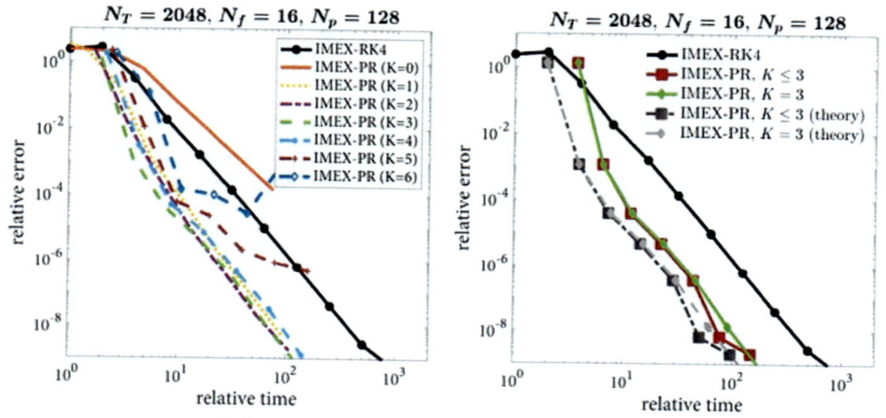

(a) (b)

Fig. 7 Relative error versus computational time for the NLS equation solved using an IMEX Parareal methods with $N_T = 2048$ and $N_p = 128$. The left plot **a** compares the theoretical running times of seven Parareal methods that each take a different number of iterations K per block. The right plot **b** compares the real-world running times of the Parareal method with $K = 3$, and a Parareal method with the adaptive controller, where $K \leq 3$. We also show the theoretical running times of the two methods in gray to highlight the losses due to communication. In both plots, the black line shows the fine integrator that is run in parallel. All times have been scaled relative to the fine integrator at the coarsest time step

Table 2 Achieved speedup (AS) and theoretical speedup (TS) for the two Parareal configurations shown in Fig. 7b

N_T	AS ($K = 3$)	TS ($K = 3$)	AS ($K \leq 3$)	TS ($K \leq 3$)
4096	8.55	16.18	8.54	16.18
8192	9.75	16.18	9.74	16.18
16384	10.25	16.18	10.55	17.01
32768	10.56	16.18	11.06	17.16
65536	10.74	16.18	11.27	17.24
131072	10.76	16.18	12.69	19.54
262144	10.98	16.18	13.57	20.61

characteristics of a Parareal method including regions of fast and slow convergence, stable regions where convergence does not occur, and regions where instabilities will eventually contaminate the method output.

By searching through a wide range of IMEX Parareal methods, we were able to identify several stable configurations that can be used to solve dispersive equations. Moreover, each of the configurations possessed the same characteristics: low iteration counts K, large block sizes N_T, and a large number of processors N_p. We also observed that the coarse integrator is the most important factor that determines whether a Parareal method is stable, and a bad choice can single-handedly lead to an unstable method regardless of the other parameters.

More broadly, we see that convergence and stability regions are highly nontrivial and depend heavily on the parameters. It is clear that one cannot arbitrarily combine coarse and fine integrators and expect to obtain a good Parareal method for solving dispersive equations. The same lesson also applies to all Parareal parameters since serious instabilities can form by arbitrarily changing the number of iterations, the block size, or the number of processors.

Finally, we remark that the analysis presented in this work can be reused to study the properties of any Runge-Kutta Parareal method on the more general partitioned Dahlquist problem that represents both dispersive and diffusive equations. However, many of the conclusions and properties that we found are specific to IMEX methods and will not hold for different method families or for different problem types.

Acknowledgements The work of Buvoli was funded by the National Science Foundation, Computational Mathematics Program DMS-2012875.

The work of Minion was supported by the U.S. Department of Energy, Office of Science, Office of Advanced Scientific Computing Research, Applied Mathematics program under contract number DE-AC02005CH11231. Parts of the simulations were performed using resources of the National Energy Research Scientific Computing Center (NERSC), a DOE Office of Science User Facility supported by the Office of Science of the U.S. Department of Energy under Contract No. DE-AC02-05CH11231.

Appendix 1: Infinity Norm of the Parareal Iteration Matrix E

Let $\mathbf{A}(\gamma)$ be the lower bidiagonal matrix

$$\mathbf{A}(\gamma) = \begin{bmatrix} 1 & & & \\ \gamma & 1 & & \\ & \ddots & \ddots & \\ & & \gamma & 1 \end{bmatrix}.$$

Lemma 1 *The inverse of* $\mathbf{A}(\gamma)$ *is given by*

$$\mathbf{A}^{-1}_{i,j}(\gamma) = \begin{cases} (-\gamma)^{i-j} & j \leq i \\ 0 & \textit{otherwise} \end{cases}$$

Proof For convenience, we temporarily drop the γ so that $\mathbf{A} = \mathbf{A}(\gamma)$, then

$$\left(\mathbf{A}\mathbf{A}^{-1}\right)_{ij} = \sum_{k=1}^{N_p+1} \mathbf{A}_{ik}\mathbf{A}_{kj}^{-1} = \begin{cases} 0 & j > i \\ \mathbf{A}_{ii}\mathbf{A}_{ii}^{-1} & i = j \\ \mathbf{A}_{ii}\mathbf{A}_{ij}^{-1} + A_{i,i-1}A_{i-1,j}^{-1} & j < i \end{cases}$$

$$= \begin{cases} 0 & j > i \\ 1 & i = j \\ (-\gamma)^{i-j} + \gamma(-\gamma)^{i-1-j} & j < i \end{cases}$$

$$= \begin{cases} 1 & i = j, \\ 0 & \text{otherwise.} \end{cases}$$

Lemma 2 *The product of* $\mathbf{A}(\omega)\mathbf{A}^{-1}(\gamma)$ *is*

$$\left(\mathbf{A}(\omega)\mathbf{A}^{-1}(\gamma)\right)_{ij} = \begin{cases} 0 & j > i \\ 1 & i = j \\ (-\gamma)^{i-j-1}(\omega - \gamma) & j < i \end{cases}$$

Proof

$$\left(\mathbf{A}(\omega)\mathbf{A}^{-1}(\gamma)\right)_{ij} = \sum_{k=1}^{N_p+1} \mathbf{A}_{ik}(\omega)\mathbf{A}_{kj}^{-1}(\gamma)$$

$$= \begin{cases} 0 & j > i \\ \mathbf{A}_{ii}(\omega)\mathbf{A}_{ii}^{-1}(\gamma) & i = j \\ \mathbf{A}_{ii}(\omega)\mathbf{A}_{ij}^{-1} + \mathbf{A}_{i,i-1}(\omega)\mathbf{A}_{i-1,j}^{-1}(\gamma) & j < i \end{cases}$$

$$= \begin{cases} 0 & j > i \\ 1 & i = j \\ (-\gamma)^{i-j} + \omega(-\gamma)^{i-1-j} & j < i \end{cases}$$

Lemma 3 *The infinity norm of the matrix* $M(\omega, \gamma) = \mathbf{I} - \mathbf{A}(\omega)\mathbf{A}^{-1}(\gamma) \in \mathbb{R}^{N_p+1, N_p+1}$ *is*

$$\|\mathbf{M}(\omega, \gamma)\|_\infty = \frac{1 - |\gamma|^{N_p}}{1 - |\gamma|}|\gamma - \omega|.$$

Proof Using Lemma 2, the jth absolute column sum of $M(\omega, \gamma)$ is

$$c_j = \sum_{k=j+1}^{N_p+1} |(-\gamma)^{k-j-1}(\omega - \gamma)| = \sum_{k=1}^{N_p-j} |(-\gamma)^k||(\omega - \gamma)|$$

It follows that $\max_j c_j = c_1$, which can be rewritten as

$$\frac{1 - |\gamma|^{N_p}}{1 - |\gamma|}|\gamma - \omega|.$$

Appendix 2: Additional Stability and Convergence Overlay Plots

Figures 8, 9, and 10 show stability and convergence overlay plots for Parareal. The following three figures show stability and convergence overlay plots for Parareal configurations with: $N_T = 2048$, IMEX-RK4 as the fine integrator, and three different coarse integrators. These additional figures supplement Fig. 3 and show the effects of changing the course integrator.

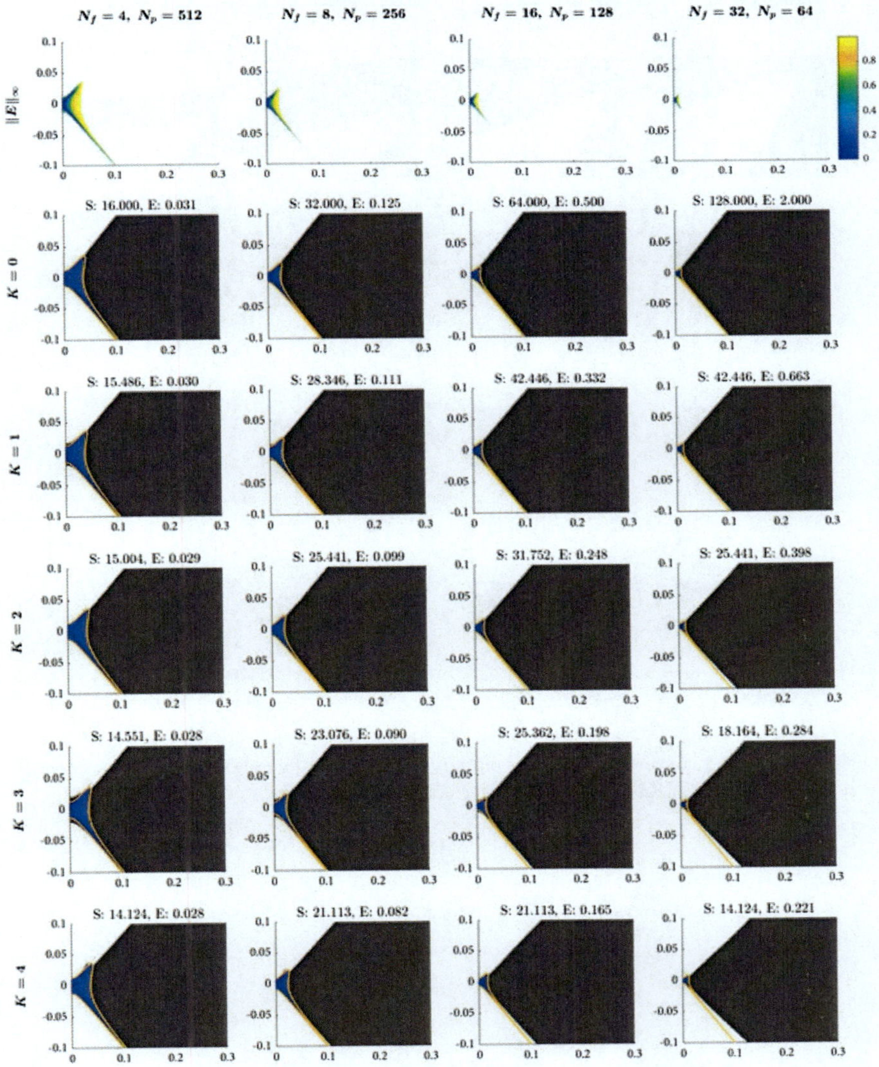

Fig. 8 Stability and convergence overlay plots for Parareal configurations with a block size of $N_T = 2048$ and IMEX-RK1, IMEX-RK4 as the coarse and fine integrators

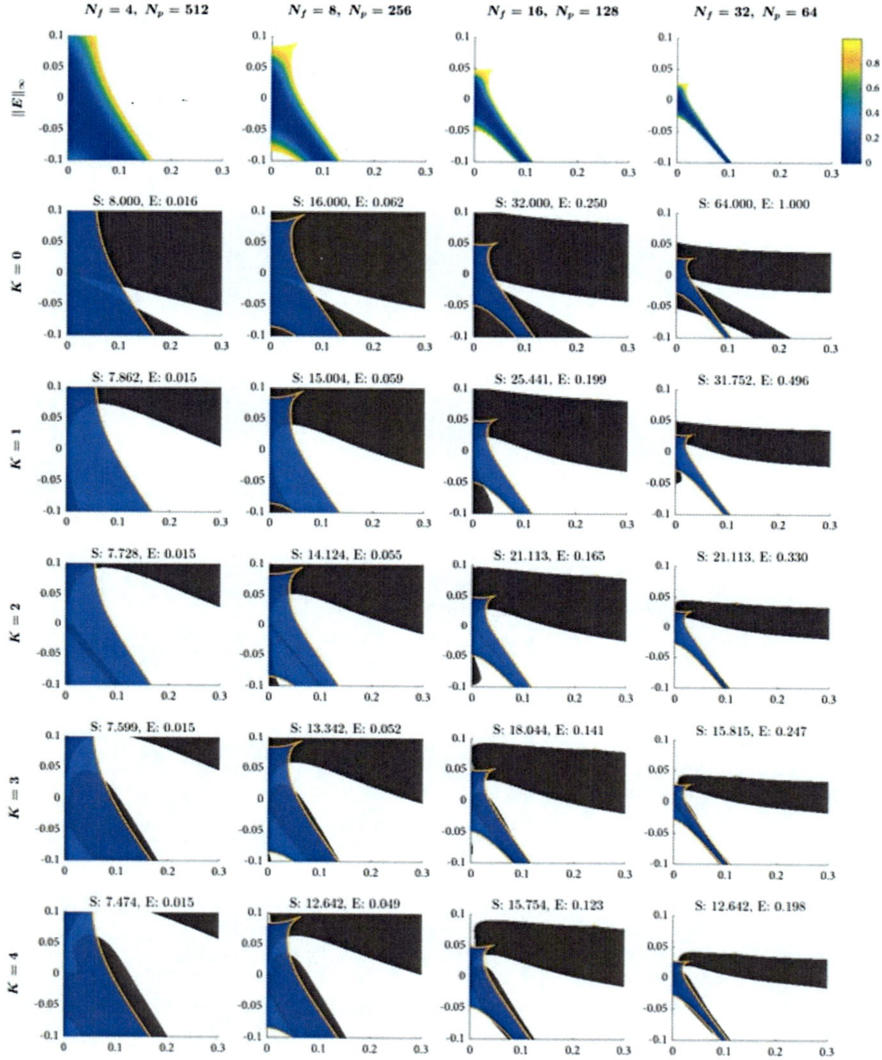

Fig. 9 Stability and convergence overlay plots for Parareal configurations with a block size of $N_T = 2048$ and IMEX-RK2, IMEX-RK4 as the coarse and fine integrators

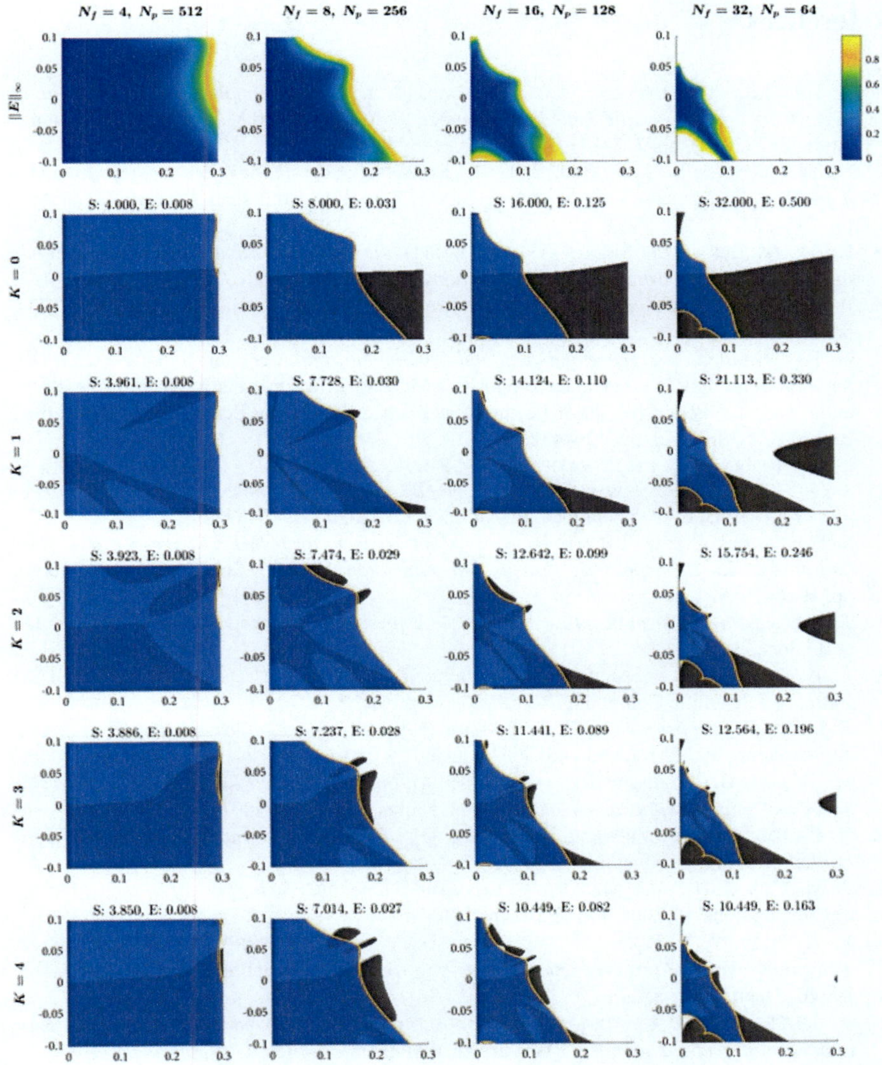

Fig. 10 Stability and convergence overlay plots for Parareal configurations with a block size of $N_T = 2048$ and IMEX-RK4, IMEX-RK4 as the coarse and fine integrators

References

1. A. Arteaga, D. Ruprecht, and R. Krause, *A stencil-based implementation of Parareal in the C++ domain specific embedded language STELLA*, Applied Mathematics and Computation, 267 (2015), pp. 727–741.
2. U. M. Ascher, S. J. Ruuth, and B. T. Wetton, *Implicit-explicit methods for time-dependent partial differential equations*, SIAM Journal on Numerical Analysis, 32 (1995), pp. 797–823.
3. U. M. Ascher, S. J. Ruuth, and R. J. Spiteri, *Implicit-Explicit Runge-Kutta methods for time-dependent partial differential equations*, Appl. Numer. Math., 25 (1997), pp. 151–167.
4. E. Aubanel, *Scheduling of tasks in the parareal algorithm*, Parallel Comput., 37 (2011), pp. 172–182.
5. G. Bal, *On the Convergence and the Stability of the Parareal Algorithm to Solve Partial Differential Equations*, in Domain Decomposition Methods in Science and Engineering, R. Kornhuber and et al., eds., vol. 40 of Lecture Notes in Computational Science and Engineering, Berlin, 2005, Springer, pp. 426–432.
6. L. A. Berry, W. R. Elwasif, J. M. Reynolds- Barredo, D. Samaddar, R. S. Snchez, and D. E. Newman, *Event-based parareal: A data-flow based implementation of parareal*, Journal of Computational Physics, 231 (2012), pp. 5945–5954.
7. S. Boscarino and G. Russo, *On a class of uniformly accurate IMEX Runge–Kutta schemes and applications to hyperbolic systems with relaxation*, SIAM J. Sci. Comput., 31 (2009), pp. 1926–1945.
8. T. Buvoli, *Rogue Waves in optics and water*, PhD thesis, Master thesis, University of Colorado at Boulder, 2013.
9. T. Buvoli, *A class of exponential integrators based on spectral deferred correction*, SIAM Journal on Scientific Computing, 42 (2020), pp. A1–A27.
10. T. Buvoli, *Codebase for "IMEX Runge-Kutta Parareal for Non- Diffusive Equations"*, (2021). https://doi.org/10.5281/zenodo.4513662.
11. M. P. Calvo, J. De Frutos, and J. Novo, *Linearly implicit Runge-Kutta methods for advection-reaction-diffusion equations*, Appl. Numer. Math., 37 (2001), pp. 535–549.
12. A. Cardone, Z. Jackiewicz, H. Zhang, and A. Sandu, *Extrapolation-based implicit-explicit general linear methods*, (2013).
13. S. M. Cox and P. C. Matthews, *Exponential time differencing for stiff systems*, Journal of Computational Physics, 176 (2002), pp. 430–455.
14. V. A. Dobrev, T. Kolev, N. A. Petersson, and J. B. Schroder, *Two-level convergence theory for multigrid reduction in time (mgrit)*, SIAM Journal on Scientific Computing, 39 (2017), pp. S501–S527.
15. A. Dutt, L. Greengard, and V. Rokhlin, *Spectral deferred correction methods for ordinary differential equations*, BIT Numerical Mathematics, 40 (2000), pp. 241–266.
16. M. J. Gander, *Analysis of the Parareal Algorithm Applied to Hyperbolic Problems using Characteristics*, Bol. Soc. Esp. Mat. Apl., 42 (2008), pp. 21–35.
17. M. J. Gander, *50 years of Time Parallel Time Integration*, in Multiple Shooting and Time Domain Decomposition, Springer, 2015.
18. M. J. Gander and S. Vandewalle, *Analysis of the Parareal Time-Parallel Time-Integration Method*, SIAM Journal on Scientific Computing, 29 (2007), pp. 556–578.
19. E. Hairer and G. Wanner, *Solving Ordinary Differential Equations II : Stiff and Differential-Algebraic Problems*, Springer Berlin Heidelberg, 1991.
20. E. Hairer, S. P. Nørsett, and G. Wanner, *Solving Ordinary Differential Equations I. Nonstiff Problems*, Math. Comput. Simul., 29 (1987), p. 447.
21. G. Izzo and Z. Jackiewicz, *Highly stable implicit–explicit Runge–Kutta methods*, Applied Numerical Mathematics, 113 (2017), pp. 71–92.
22. C. A. Kennedy and M. H. Carpenter, *Additive Runge-Kutta schemes for convection-diffusion-reaction equations*, Appl. Numer. Math., 44 (2003), pp. 139–181.

23. S. Krogstad, *Generalized integrating factor methods for stiff PDEs*, Journal of Computational Physics, 203 (2005), pp. 72–88.
24. J.-L. Lions, Y. Maday, and G. Turinici, *A "parareal" in time discretization of PDE's*, Comptes Rendus de l'Acadmie des Sciences - Series I - Mathematics, 332 (2001), pp. 661–668.
25. M. Minion, *Semi-implicit spectral deferred correction methods for ordinary differential equations*, Communications in Mathematical Sciences, 1 (2003), pp. 471–500.
26. M. L. Minion, *A Hybrid Parareal Spectral Deferred Corrections Method*, Communications in Applied Mathematics and Computational Science, 5 (2010), pp. 265–301.
27. J. Nievergelt, *Parallel Methods for Integrating Ordinary Differential Equations*, Commun. ACM, 7 (1964), pp. 731–733.
28. D. Ruprecht, *Shared Memory Pipelined Parareal*, Springer International Publishing, 2017, pp. 669–681.
29. D. Ruprecht, *Wave Propagation Characteristics of Parareal*, Computing and Visualization in Science, 19 (2018), pp. 1–17.
30. A. Sandu and M. Günther, *A generalized-structure approach to additive Runge–Kutta methods*, SIAM Journal on Numerical Analysis, 53 (2015), pp. 17–42.
31. B. S. Southworth, *Necessary Conditions and Tight Two-level Convergence Bounds for Parareal and Multigrid Reduction in Time*, SIAM J. Matrix Anal. Appl., 40 (2019), pp. 564–608.
32. B. S. Southworth, W. Mitchell, A. Hessenthaler, and F. Danieli, *Tight two-level convergence of linear parareal and mgrit: Extensions and implications in practice*, arXiv preprint arXiv:2010.11879, (2020).
33. G. A. Staff and E. M. Rnquist, *Stability of the parareal algorithm*, in Domain Decomposition Methods in Science and Engineering, R. Kornhuber and et al., eds., vol. 40 of Lecture Notes in Computational Science and Engineering, Berlin, 2005, Springer, pp. 449–456.
34. Z. Wang and S.-L. Wu, *Parareal algorithms implemented with IMEX runge-kutta methods*, Mathematical Problems in Engineering, 2015 (2015).
35. G. Wanner and E. Hairer, *Solving ordinary differential equations II*, Springer Berlin Heidelberg, 1996.